iPHONE SENIORS GUIDE

The Most Intuitive and Exhaustive Manual to Master Your New iPhone from Scratch.
A Detailed Step-by-Step Guide with Pictures and Useful Tips & Tricks

George Halperin

TABLE OF CONTENTS

INTRODUCTION

Before getting too excited about another Apple iPhone model release, make sure you know everything there is to learn about your iPhone 13. Apple is a company that seems to always have something up in its sleeve.

However, iPhone models do not come cheap, so it is best to make the most out of your phone before changing it to something more advanced.

If you are an iOS user or want to try Apple devices, you are probably wondering which product line to consider. Apple devices come in a wide range of product lines, with the iPhone 13 being among the latest iPhone generations. This 2021 model has left many wondering whether it is a worthwhile investment from Apple's 2020 release.

But what exactly has changed about this model, and is it worth the investment? Although the iPhone 13 series is akin to the iPhone 12 generation, it has new features and design tweaks that give it a powerful performance.

Find out more about your iPhone 13 by exploring all its features and understanding the phone through and through. You can also use this to guide you in making an informed buying decision on whether or not it's time for an upgrade, you better wait for the next one, or you're more compatible with the old iPhone model you're using.

CHAPTER 1: TERMINOLOGY

The iPhone 13 is an addition to the many iPhone models, which Apple introduced in September 2021. True to the previous product lines, it combines the technologies of a cellular phone with a touch screen interface, digital camera, iPod, and a computer all in one. All phones run using the iOS operating system.

The iPhone 13 is available in four models - the iPhone 13 Pro Max, iPhone 13 Pro, iPhone 13 mini, and the base iPhone 13. All these models have integrated a 6-core CPU known as the A15 Bionic chip, which is a proprietary of the brand.

Here are the basic terminologies common in all iPhone 13 models:

Bluetooth 5

You may think that it's only Bluetooth; most of your devices have it, so what's the big deal with this version? The evolution included crucial improvements that allowed operation modes to use low energy. The wireless protocol has gotten stronger and better; it can now link your phone to wearables, heart monitors, weather stations, headphones, and more. This is only the beginning, and more devices are expected to hook up as the technology continues to evolve.

It makes the connection and transfer of files between Apple devices faster and more convenient. The speed has doubled from the Bluetooth 4 to 4.2 with up to 1Mbps throughput. Bluetooth 5 is capable of speeds up to 2Mbps. This allows a higher bitrate streaming for audio and a faster connection with smartwatches.

For now, Bluetooth 5 lays the foundation for seamless wireless streaming for future devices since Apple still hasn't announced plans to utilize higher bitrate streaming. The iOS supports 256 Kbps max bitrate of Advanced Audio Coding (AAC), and Apple Music currently streams at the same speed.

However, Apple will likely use the technology to allow a single iOS device to hook multiple Bluetooth audio devices. Moreover, the technology is currently being prepared to connect the growing numbers of the Internet of Things (IoT). The name is Bluetooth 5, which, unlike the previous versions, doesn't have an LE, v, or a decimal point. The reason for the name is to make it easier for the public to understand it. Before it came to this, Bluetooth LE was introduced as a low-energy protocol, which was something new at the time of its release. Certain models of the iPhone utilized the lower versions of Bluetooth as well.

Additionally, when it comes to range, Bluetooth 5 can reach up to 240 meters or 800 feet range. Many users would not mind or be aware of the range until the audio being played stops or starts becoming choppy, which are signs that you have reached the maximum range the wireless connection is capable of. It is important to note that aside from range, there are obstructions that may block Bluetooth's range and connection, including walls.

This Bluetooth version also has higher maximum bandwidth. Its new physical layer (PHY) has more than 20 dB transmission power in low energy mode and supports 2 megabits per second of speed. This means that Bluetooth 5 has two interfaces used when operating with low energy - it transmits double the data despite the shorter range and transmits lower data at longer distances.

Aside from the iPhone 13 models, Bluetooth 5 has been integrated into other Apple devices, such as the MacBook Pro 2018 with Touch Bar, HomePod, iPhone X, iPhone 8 Plus, and iPhone 8. Apple was among the first phone companies to ship phone models compatible with this version of Bluetooth. While it is quickly becoming common in the more advanced iPhone models, including iPhone 13 and other high-end phones, this Bluetooth version needs to get adapted to more devices to use your iPhone 13 to connect with more gadgets.

The good thing about Bluetooth 5 is that you can connect it to older Bluetooth versions. This means that iPhone 13 will work even with devices using prior Bluetooth versions, including automobiles, fitness trackers, speakers, Bluetooth headphones, and more.

Wi-Fi 6 (802.11ax)

All iPhone 13 models are equipped with Wi-Fi 6 or what was previously called 802.11ax. Renaming Wi-Fi models to something easier to remember was pushed by the Wi-Fi Alliance since previous names, aside from 802.11ax, include 802.11af, 802.11ac, 802.11n, and more. The predecessor, Wi-Fi 5, has only been named such in October 2018, so the term is rarely used.
Wi-Fi 6, or the next generation of Wi-Fi, is faster, and people will continue to witness its benefits through time. It is designed for a future-facing upgrade. To give you an idea about the boost in terms of speed, Wi-Fi 5 is capable of 3.5 Gbps while it's 9.6 Gbps with Wi-Fi 6. You can split the connection to multiple devices, giving each gadget a higher potential speed.

The technology allows all connected devices to improve their network and speed, and it is not solely for boosting the speed of a solo gadget. At the time of Wi-Fi 5, an average home had at least five devices connected to each Wi-Fi modem. The numbers have increased to nine in recent times, and it is said that after several years, it will hit the mark of up to 50.

That is the number of devices you can allow to connect to the Wi-Fi on your iPhone 13 when you share the connection or use it as a Hotspot as long as it's connected to a Wi-Fi 6 router. If you want to know how fast is fast, you can only conclude that it's significantly quicker. The speed of connection for each gadget using your phone as a Hotspot will vary depending on the devices used. However, it will be easier to differentiate when more gadgets start connecting to Wi-Fi 6 since it will retain its stable speed compared to the same number of devices connecting to a Wi-Fi 5.

Aside from iPhone 13, newer devices will begin coming up with versions using Wi-Fi 6 as a default connection. Despite being up-to-date, you can only benefit from the fast internet connection of the iPhone 13 if you connect it to a Wi-Fi 6 router. Another good thing with Wi-Fi 6 is that it requires minimal battery from the connected gadgets. This is possible through the Target Wake Time feature that allows routers to work with the connected devices' check-in times. It is also made with better security features.

5G chipset

It's an improved network infrastructure and customer premises equipment designed to enable users to build a wireless network based on the specification of a 5G-network. The term 5G pertains to fifth-generation wireless, the newest of its kind after the introduction of 4G LTE. Both networks allow global linking of mobile internet to the IoT or Internet of Things.

IP68-rating

The iPhone models use the IP or ingress protection rating of two digits. This has been implemented since the release of the iPhone 7. The first digit of the rating ranges from zero to six measures the device's protection against solids, including dirt and dust. The second digit, which ranges from zero to eight, measures the gadget's protection against liquids. When the device gets a rating with eight as its second digit, it can remain submerged in water deeper than one meter, but only with the exact time of submersion and distance provided by the manufacturer.

The more water-resistant devices now have an IP rating of 9K, which means that a phone with an IP69K rating can last longer submerged in water than a device with an IP68 rating. However, only a few gadgets have this rating at the moment, and it might take years before it can be implemented in many phone models.

This might leave you wondering - if the iPhone 13 has an IP68 rating, does it mean that it's safe to use underwater? You can at your own risk. However, it is not recommended by its manufacturer. Apple has acquired a 2019 patent for using their phone's camera underwater, but the technology has yet to be perfected. The patent pertains to the device's capability when it is being used underwater due to a pressure sensor.

All iPhone 13 models have undergone controlled laboratory testing to test their resistance against dust, water, and splash to get the rating. They got an IP68 rating, meaning that the phone models can be submerged underwater for up to 30 minutes with a maximum depth of 6 meters. Remember that your phone's resistance will wear out in time, and Apple is not liable for damages incurred by the phone due to intentional submersion to liquids.

However, it is safe to say that all the iPhone 13 models are resistant to accidental liquid spills. When you spill your phone with juice, tea, coffee, beer, or soda, you can safely rinse the spills with tap water and dry off the phone's surface by wiping it with a clean cloth.

Here are the things you have to avoid encountering when using any iPhone 13 model to prevent liquid damage:

- Removing the phone's screws and trying to disassemble the device
- Dropping the phone
- Using it in extremely hot or cold weather conditions

- Submerging the phone in water for long hours
- Using the gadget in a steam room or sauna
- Exposing the phone to high velocity or pressurized water
- Bathing or swimming with your phone

Cinematic mode

You will find this feature inside the Camera app. It's like a Portrait mode but a video version. You can play with the effects once you are in this mode. You can adjust the focus and choose which you will blur or retain the clarity - the background or foreground. It also allows you to use blurred and soft bokeh effects or make the image crisp by applying a depth-of-field effect. It makes shooting flexible with an output like it was shot by a pro.

Another good thing about the Cinematic mode is that it doesn't slow down the phone's operation. It runs inside the A15 Bionic Chip, which third-party applications can't access. This mode tracks multiple points on the subject while you are shooting, so you can follow more than one as the object of focus.

Your phone allows the perfect captures of the entries and exits as you move the camera and naturally change the focus. The result is picture-perfect since the phone is equipped with ultrawide lenses. Your iPhone will also save your preference as you use the cam, so it's easy to change the focus to a new subject if ever you change your mind.

OLED Super Retina XDR

This is the display feature available to all iPhone 13 models. The display extends to the chassis and is pretty flexible. The phone lineup has a blacker black to brighter white ratio of 2,000,000:1. They have a peak brightness of 1200 nits for movies, TV shows, videos, and HDR photos. Apple claims its phones to have a 25 percent brighter display when used outdoors. The typical maximum brightness for the standard iPhone 13 models is 800 nits, and for the Pro models, it's 1000 nits.

Among all iPhone 13 models, many reviews have given the iPhone 13 Pro Max commendation for having the best display, especially when it comes to contrast ratio, color accuracy, max brightness, and HDR. Additionally, the display is responsive due to the phone's support for Haptic Touch and oleophobic coating that is fingerprint-resistant. The phones have an integrated True Tone, which makes them easy on the eyes, and the wide color support offers real-life and vivid colors.

Here are the other terminologies used in all iPhones and other Apple gadgets you should know about:

iOS

If you are familiar with Windows, the operating system of many PCs, this is its counterpart on iPhones. The iOS used to be called iPhone OS back in 2007 when the first iPhone was introduced. Later on, with the iPhone 3G, the operating system became iOS. The version number commonly follows the term. For example, iPhone 13 models have an operating system of iOS 15, which you can choose to upgrade up to version 15.4.1.

Siri

Your phone's virtual assistant makes it easier to use the device or create entries to notes or calendars. You can also ask Siri about anything, which can be entertaining at times. Siri also makes it possible to control your smart gadgets at home through your phone. You can also monitor what's going inside while away and ask Siri to perform different tasks, such as turning the music off or turning on the lights. You can start talking to Siri by pressing and holding the Home button.

iCloud

This is the umbrella term used by Apple for its online-based data services and technologies, including iCloud.com, iCloud Backup, iCloud Keychain, and iCloud Drive. You can buy more storage if you intend to use the services to store more data.

Personal Hotspot

This feature lets you share your internet connection with other gadgets via different options, including USB, Bluetooth, or Wi-Fi. You can turn this feature on under Personal Hotspot, which you can see upon clicking Settings.

Apple Pay

You can add a debit or credit card to your iOS Wallet app. Once you have set this up, you can use your phone to pay for purchases at the point of sale terminals. However, you must first check the feature's availability in the area where you are based before setting it up.

Control Center

You can customize the Control Center to make it easier to access and use your apps by swiping up from the screen's bottom.

Face ID/Touch ID

They are used to authenticate the biometrics of the phone's owner. You can choose between Face ID or facial recognition and Touch ID or your fingerprint. Once the phone is unlocked, you can continue using it, play media, purchase items, and do whatever you like using your phone. You will be prompted to set this up if it's your first time using them and choose to disable the features whenever preferred.

AirDrop

This feature appears when you click Share to transfer files locally. It allows you to send the files to other iPhone users that appear in your list of recipients. You can also open or close your AirDrop visibility.

CHAPTER 2: MOST IMPORTANT THINGS TO KNOW

Types of iPhones

There are four types of iPhone 13 models on the market, and each varies in price, color, battery capacity, storage, and size. Apple's iPhone 13 was first launched in September 2021 and features models like iPhone 13 mini, iPhone 13, iPhone 13 Pro, and iPhone Pro Max.

Major Differences

Each device in iPhone 13 product line differs in some ways. Here is a detailed comparison of the 4 models that you may find worthwhile knowing when making your decision.

1. **Battery Life**

A major difference between the iPhone 13's line is in the battery life. The iPhone 13 mini can last up to 13 hours of video streaming while the iPhone 13 gives up to 15 hours of video streaming. Although the iPhone 13 Pro is similar in size to iPhone 13, it has more battery life of up to 20 hours of live streaming. On the other hand, the iPhone 13 Pro Max has the largest battery life, lasting up to 25 hours of video streaming.

2. **Storage**

Another major difference between the iPhone 13 models is their storage capacity. Unlike the standard 64GB capacity options, Apple has upgraded its storage capacity options to 128GB, 256GB, and 512GB for the iPhone 13 and iPhone 13 mini options.

Besides this, Apple offers an additional 1TB storage option for the Pro and Pro Max models. Impressively, this is the first time iOS users get to enjoy 1TB of storage on their iPhones. Of course, the price for each model varies depending on the amount of storage.

3. **Display**

It is also the first time that Apple phones, specifically iPhone 13 Pro and iPhone 13 Pro Max are equipped with a 120 Hz refresh rate. Although this feature is already available in high-end models on the market, it is an upgrade from the usual 60Hz refresh rate on previous iPhones.

It entails the number of frames that a screen can display per second when one is scrolling the web pages and apps. Generally, the higher the hertz number, the fast and smooth things will appear on the screen. This feature will make your phone feel faster and more powerful and look more premium.

4. **Camera Upgrades**

Each model in the iPhone 13's line features new camera upgrades that were not available in the previous line-up. This includes features like Cinematic Mode, which is available on iPhone 13 and helps to automatically shift the focus when the subject changes the direction or someone else gets into the camera frame.

The iPhone 13 and 13 mini models are designed with a new dual-camera system that has a wide camera. This camera system facilitates low light performance thanks to the fact that it allows up to 47% more light. What's more, this feature helps to ensure noise reduction.

The Pro and Pro Max models feature a lidar scanner feature, which is usually available in iPhone's high-end models. This feature comes in handy when taking pictures in low light. It is also a great autofocus tool and it can be used to measure the subject's height.

Also, like the Pro Max model, the iPhone 13 Pro now comes with the increased night mode performance feature. Additionally, this model is an excellent choice for shooting professional videos and lets you make live photo edits during shooting, thanks to its Photographic Styles System.

Similarities

Despite their differences, the iPhone 13's line of models features similar OLED displays, square designs, iOS software, A15 chip, IP68 water, and dust resistance, and 5G compatibility. It is also worth noting that they all lack expandable storage and a headphone jack.

SOS Mode

Emergency SOS is a software feature integrated into iPhone to help you quickly and easily alert your emergency contacts and call for help. This is a useful safety feature that can come in handy when you are in trouble and you need to alert your emergency contacts to your whereabouts.
You can set up and activate this feature on your iPhone in several ways. They include:

Step 1: Quickly Press the Home/Side Button Five Times

Apple provides the option to press the home or side button on your device to enable the SOS mode. To activate this function, you need to:

- Open the Settings menu
- Go to Emergency SOS
- Locate the Call with Side button and turn it on

Step 2: Press and Hold the Home Button and the Volume Up/Down Button

When you press and hold the home button and the volume up or down button, a screen will pop up giving you several options.

One of the options is to slide for Emergency SOS, alerting your emergency contacts. However, it is worth noting the slide option may not be very convenient as the first method.

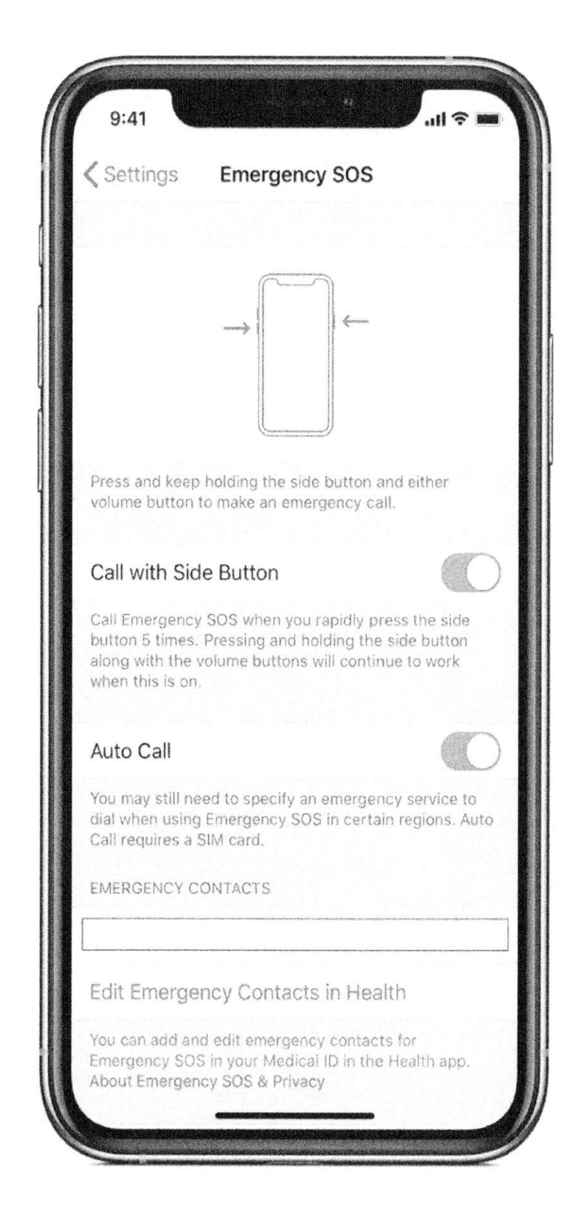

What to Do When Your Phone Gets Stuck on Emergency SOS Mode

A common issue that most iOS users encounter is their device getting stuck on emergency SOS mode. This can be a frustrating issue since the emergency SOS screen may not go away, making it nearly impossible to use the phone. Usually, this can occur because of making wrong operations like rooting your iPhone, pressing the power button with the volume up or down, etc.

The good news is that this issue can be fixed using different methods despite the cause. For instance, you can use an iOS repair tool to reset your device in one click without losing any data.

The other method is to force restart your phone by quickly holding and releasing the volume up and down button and then pressing the side button for several seconds until you see the brand logo.

You may also want to try updating your iPhone in recovery mode to resolve the issue of SOS mode. If all else fails, consider contacting Apple support to help resolve the iPhone stuck in the SOS mode issue.

Siri

Looking for a quick way to get things done on your iPhone when you can't get your hands on it? Well, talking to Siri can make this possible. Like other iPhone models, this functionality is useful for iPhone 13 users because it can help you do multiple things like find a location, translate a word, sort our schedules, and much more.

To use this functionality, you need to set it up on your iPhone 13. To do this:

Activate Siri with a Button

- To access Siri using a button, go to Settings then **Siri & Search**.
- Turn on the "**Press side button for Siri**."

If your phone is in silent mode, Siri will respond silently. It will, however, respond loudly if the silent mode is turned off.

Activate Using Your Voice

If you prefer to activate it using your voice, turn on the **Listen for "Hey Siri"** toggle. In this case, your iPhone will listen for voice commands instantly. To ask a question or give a command, you can access Siri by saying '**Hey Siri**.' Then proceed by asking a question or the task that you would want Siri to do for you.

NB: If you don't want to use the Siri voice control in some instances, be sure to place your device face down. You can also turn off the functionality by going to **Settings** and then clicking on **Siri & Search**. Toggle off the **Listen for "Hey Siri**."

Typing Instead of Speaking

If you prefer to type to Siri rather than talking, you can customize the process using accessibility controls. To do this:

- Go to **Settings** and then click on **Accessibility**.
- Click on **Siri** and turn on the **Type to Siri** option.

CHAPTER 3: HOW TO SETUP YOUR IPHONE

Language Setup

Usually, you are allowed to change the language and region of your device during the initial setup. However, one can accidentally change the language on their iPhone 13 to a language that they don't understand.

In other instances, your iPhone can come in a different language, especially if it's second-hand or imported. You may also need to change the language when you are traveling or moving to another region.

Follow the step-by-step instructions below for changing the language on your Apple device.

Step 1: On the Home screen, go to the Settings menu, which has an icon of a cogwheel.

Step 2: From the list provided, navigate to the General option, which also has the icon of a cogwheel.

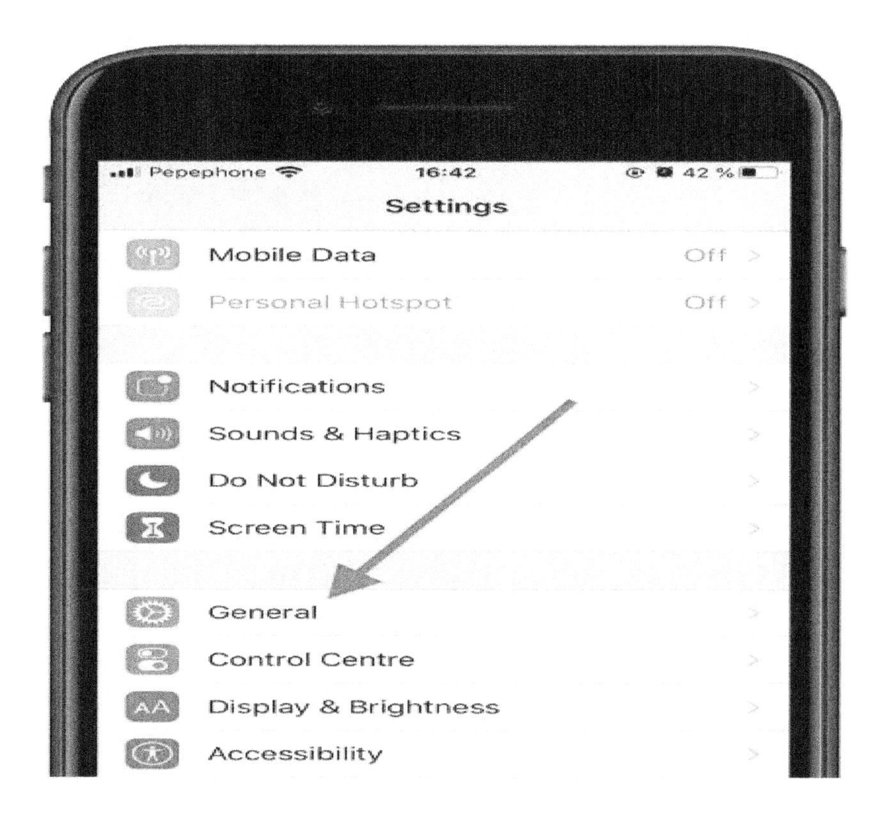

Step 3: Navigate to the **Language & Region** section and tap on it.

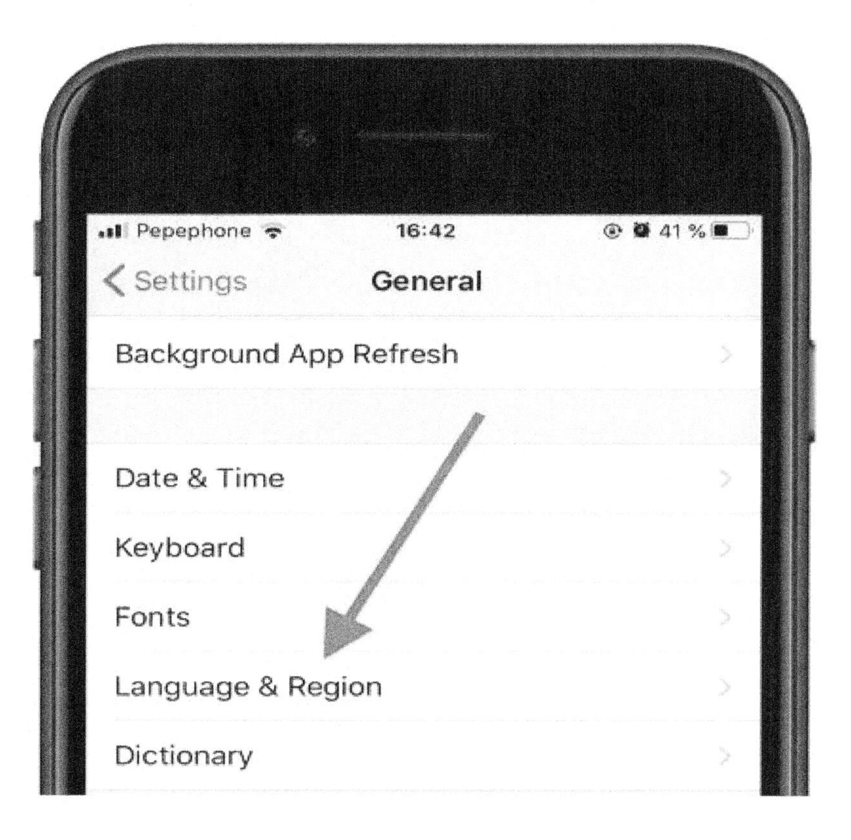

Step 4: At the top option, click on **iPhone Language** to see the language that is currently installed on your device. You will also see a list of some of the suggested languages.

Step 5: Click on the language of your choice. However, if you can't find your preferred language on the list provided, use the search option at the top to look for it.

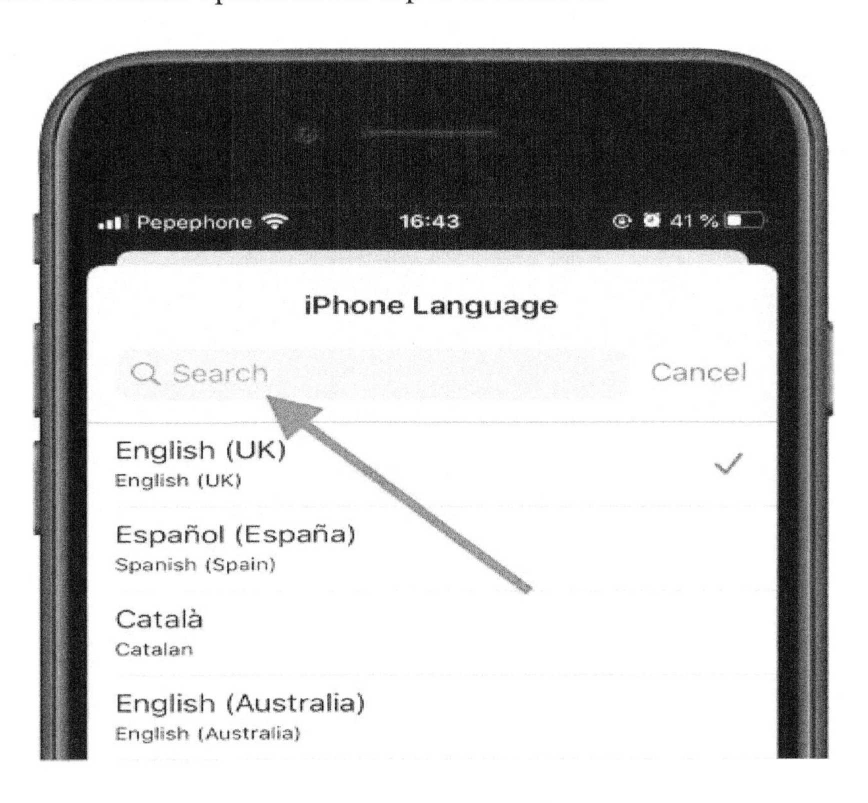

Step 6: The next step is to confirm your selection. You will get an alert asking you to confirm the change to the new language. To accept, click **Continue** or the first option provided.

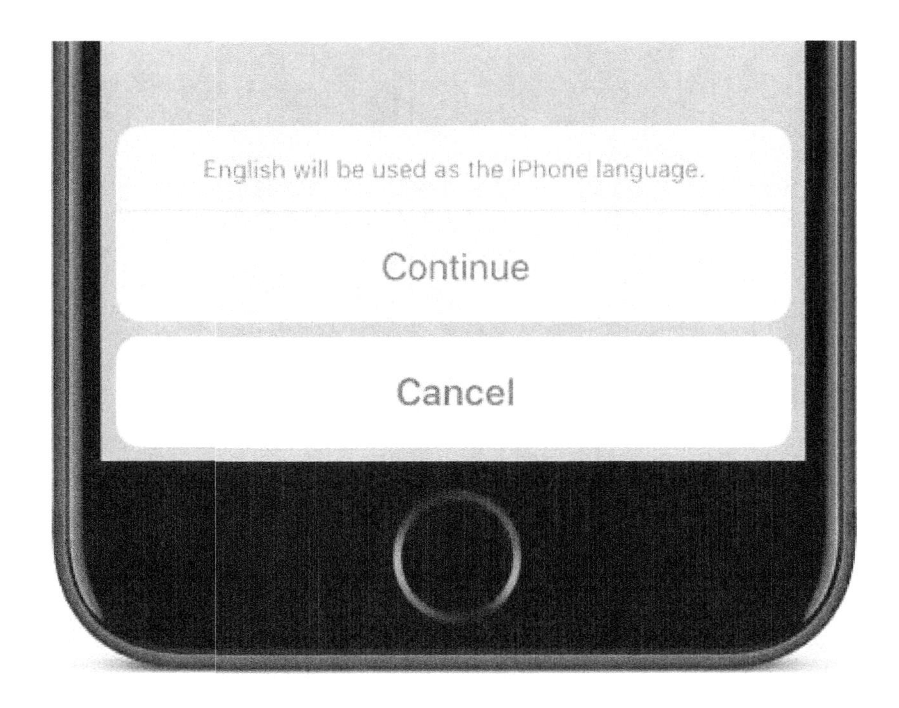

Step 7: After confirming your selection, your selected language will appear as the default language for your iPhone 13.

Step 8: If you want to add another language, click on the **Add Language** option and then choose the language of your choice.

Once you have tweaked your iPhone language, you will now be able to use all the settings using that language. This includes the functions and instructions.

Setup Process

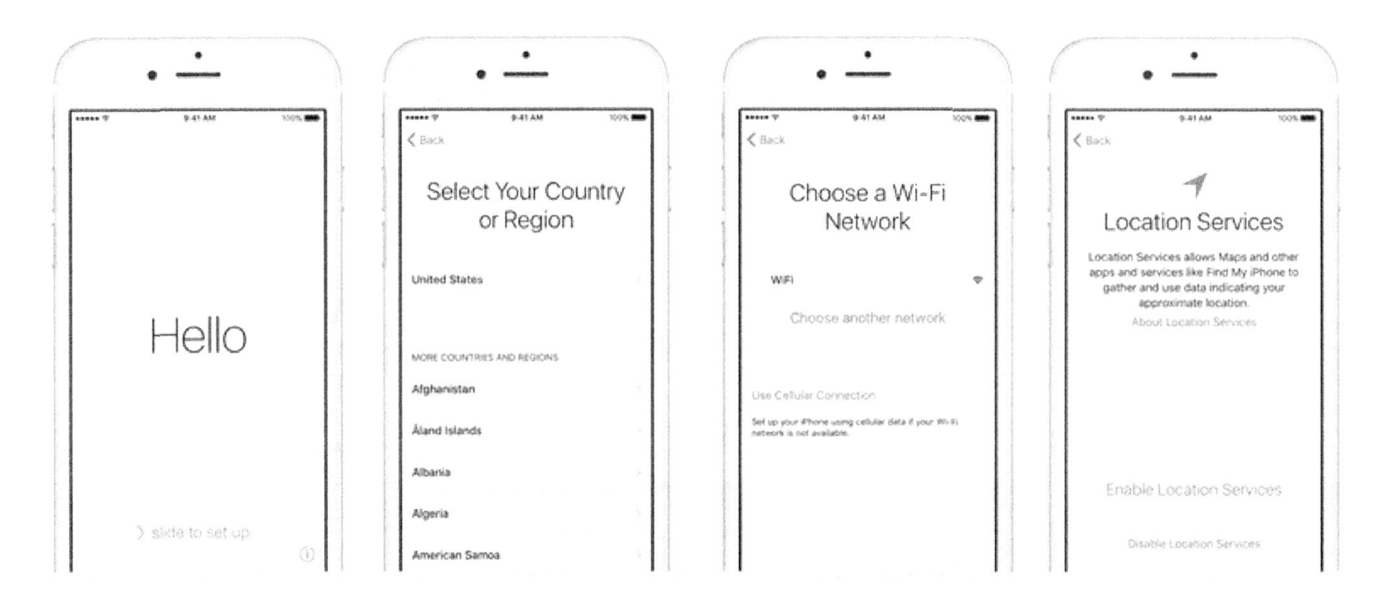

Once you unbox your new iPhone 13, the next step is to set it up to keep it up and running. However, for most people, this can be a daunting task especially if you already have an existing iPhone and want to transfer everything to your new iPhone. Whether you are setting up your iPhone for the first time or after doing a factory reset, this guide will come in handy.

Back Up Your Old Device

Before you get started with the setup process for your new iPhone, make sure that you back up your old iPhone so that you don't lose your data. There are several ways in which you can back up your iPhone including using iCloud, a PC, or a Mac.

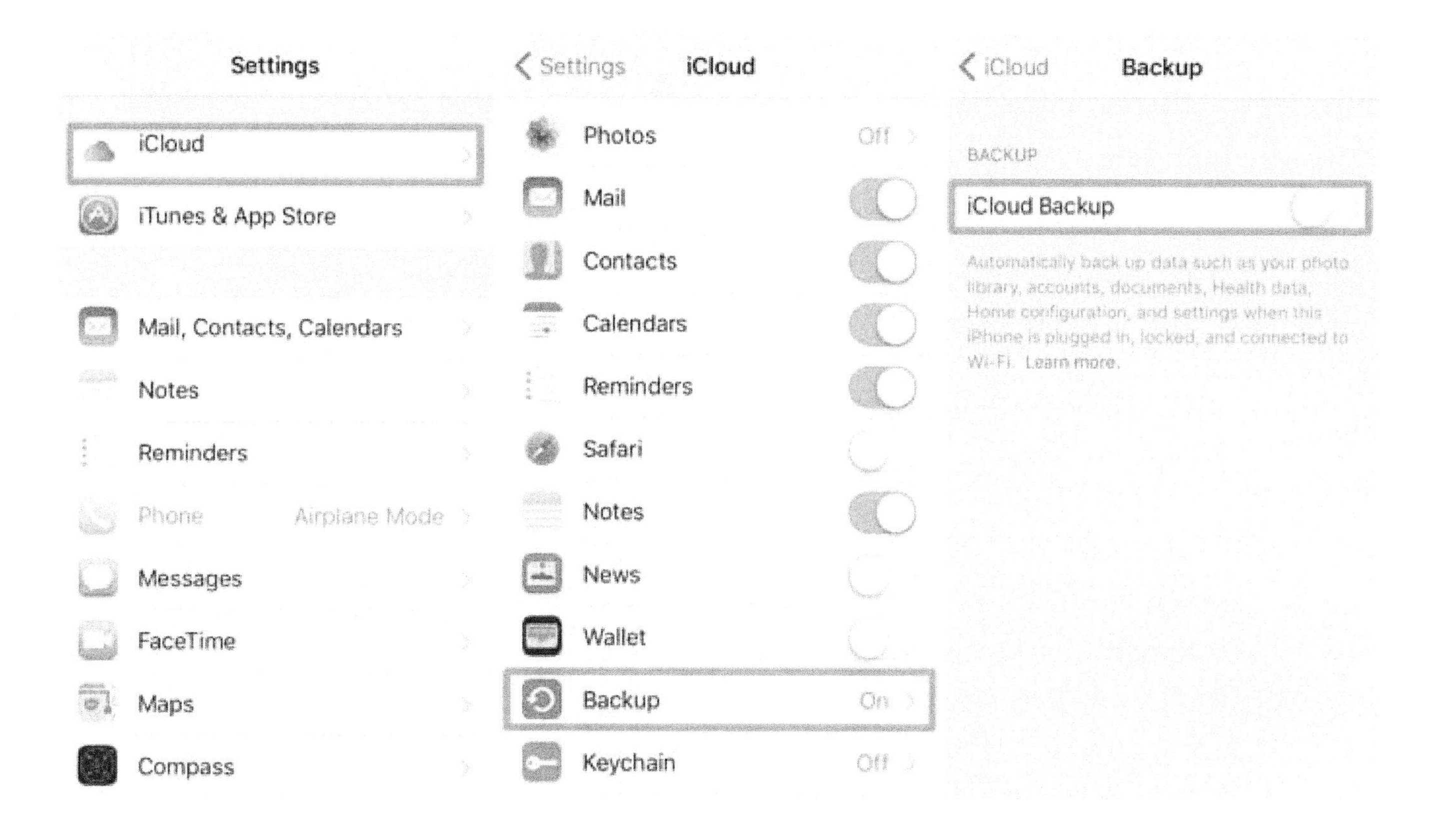

Turn On Your New iPhone

Step 1: Now that all your data is backed up, turn on your new iPhone device. To do this, press and hold the side button until you see the Apple logo.

Step 2: The "**Hello**" phrase will appear on your screen in different languages regardless of the method you choose to set up your iPhone.

Step 3: On the bottom part of the screen, you will see the option to Swipe to get started. However, if you are using an iPhone SE, you will need to click the Home button to get started.

Step 4: Choose your **language** from the options provided.

Step 5: Next, you need to choose your **country or region** from the list provided.

Step 6: At this point, you will be given the option to choose between the option for QuickStart setup and the option to set up manually. The QuickStart setup option may be worth considering if you want to automatically transfer data from your old phone.

Step 7: If you choose the Set Up Manually option, you will be prompted to choose a Wi-Fi network. If you are unable to set this up at the moment, opt for **the Use Cellular Connection** option.

Step 8: Read the Data & Privacy information provided and then click on **Continue**. If you still need more insight, select **Learn More.**

Step 9:

Tap Continue and you will be prompted to set up Face ID, where you will use your face for biometric security. You can do this now or skip it by selecting "**Set Up Later**." If you choose to set up Face ID, you will need to enter a passcode and re-enter it to confirm. If you don't want to create a passcode, you can skip this process by selecting the **Passcode Options** and then tapping on **Don't Use Passcode**.

Step 10: In the next step, you will be provided with several ways to set up your device. They include:

- **New Set-Up**

As the name may suggest, this is setting up your iPhone right from scratch. You may consider this setup option if you want your iPhone to feel new or if it is your first time using a smartphone.

- **Restore from Previous iPhone**

If you've owned a previous iOS device, you can set up your iPhone through the internet using iCloud. Alternatively, you can also transfer all your data via USB with iTunes to your new iPhone device.

- **Transfer Directly From iPhone**

If you already own an old iPhone device, Apple provides the option to transfer data from your old device to your new device. For this, you can use the Quick Start Option to quickly and safely move apps and settings from your previous device to your new iPhone device.

- **Move Data from Android**

The other option for setting up your device is importing your data from Android or any other mobile device. This method is for anyone that had a previous mobile device and is now switching to an iPhone or iPad. Apple makes this possible through an app available in Google Play.

Before you restore data from iTunes, iCloud, or from an old device, ensure that you first backup your data.

Face ID

Like other iPhone product lines since the iPhone X, iPhone 13 uses the face ID to unlock and authorize App Store purchases. This biometric security can also come in handy when one is signing in to third-party apps.

This feature is considered very secure because it reads the entire face. It requires you to show all angles of your face by moving your head in a circle. Nonetheless, Apple provides the Accessibility option for those with physical limitations and who can't perform a full head motion, have low vision, or are blind.

How to Set Up Face ID

- Click on the **Settings** icons and locate **Face ID & Passcode**.
- Navigate to Set Up Face ID and click on it.
- On the How to **Set Up Face ID** option, click on Get Started.
- Follow the instructions and click **Done**.
- At this point, you can set up a phone lock code and confirm it.
- Still, on the Face ID & Passcode section, you will see a list of options for which you want to use the face ID. Turn on or off the toggle depending on which settings you want to use the Face ID.
- Once done, return to the home screen by sliding your fingers upwards from the bottom of the screen.

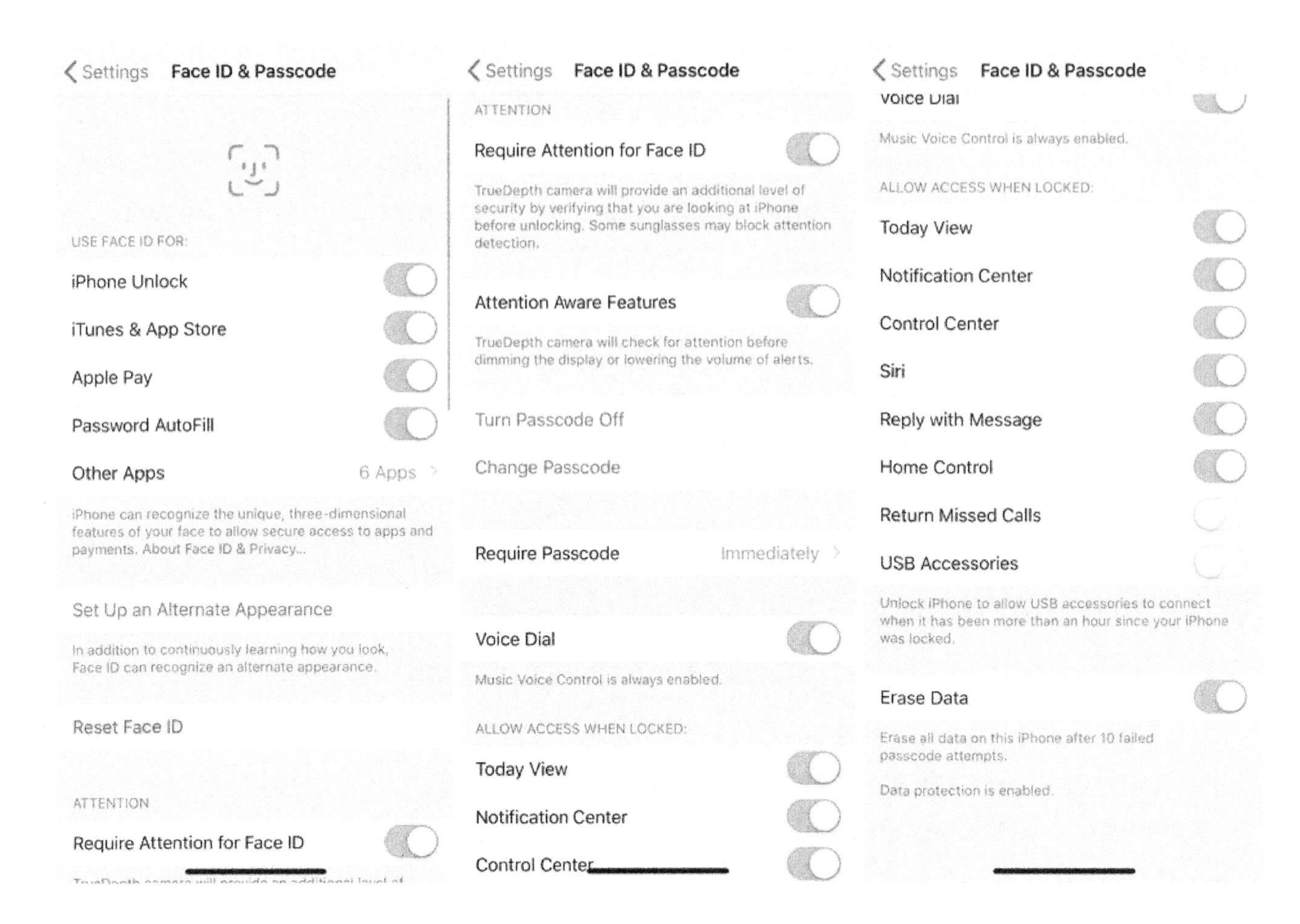

How to Set Up Accessibility Options

If you can't move your head in a full range motion, you can set up your Face ID with accessibility options. Here is how to:

- Open **Settings**.
- Go to **Face ID & Passcode**.
- Place your face in the frame.
- Click on **Accessibility Options**.

How to Change Attention Settings

The Face ID in iPhone devices is attention aware, meaning that it can only unlock your device when your eyes are opened and focused on the screen. This helps to provide additional security. It can also ensure that the screen is well lit when your eyes are focused on the screen and much more.

To turn on or off this feature:

- Go to **Settings**.
- Tap on **Face ID & Passcode**.

- Toggle on or off any of these settings:
    - ✓ Attention Aware Feature
    - ✓ Require Attention for Face ID
    - ✓ Haptic on Successful Authentication

Apple Pay

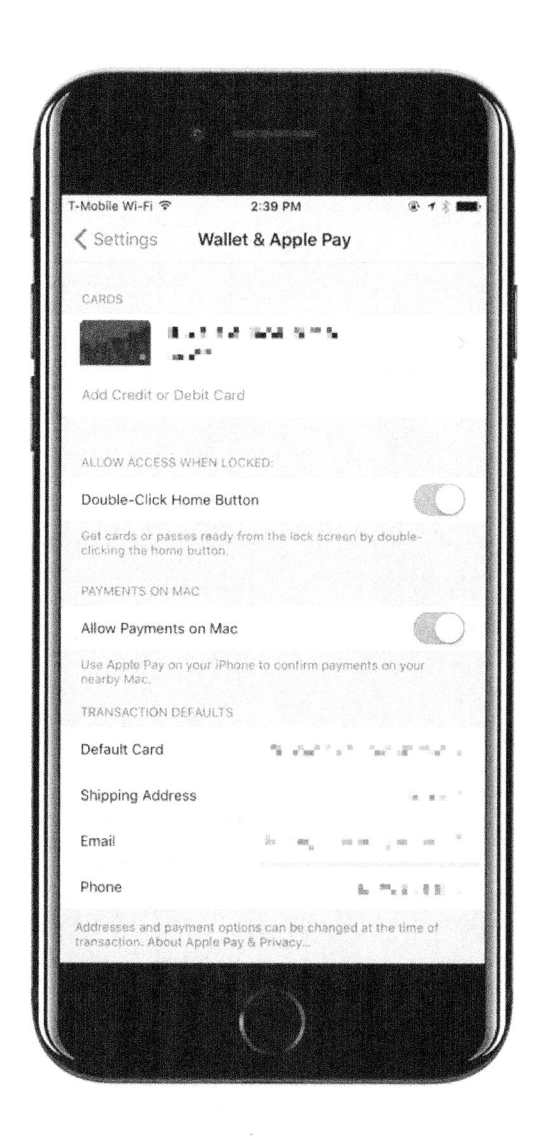

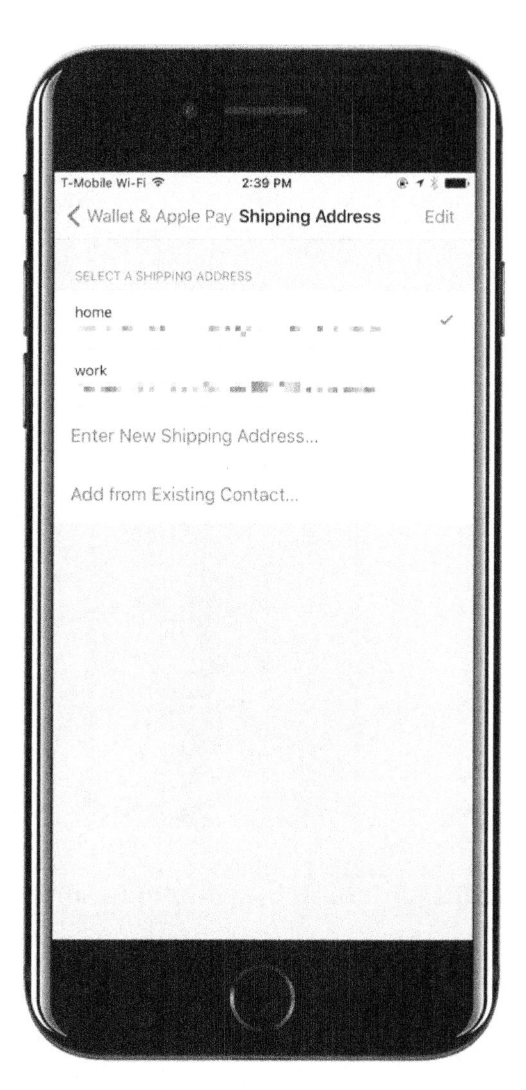

Besides setting up Face ID in your new iPhone device, you also need to set up Apple Pay. Apple Pay is built into Apple devices and is a simple and safe way to make payments in stores, websites, apps, and much more, as long they support it.

To set up this feature in your iPhone 13, you will need to add your credit, debit, and prepaid card to your wallet. To add a card, follow these steps below:

- Go to wallet and tap on + on the upper right side.

- Next, you will be prompted to sign in using your Apple ID.
- Once logged in, you can add a debit or credit card. Place the card in your phone's frame to scan the details or enter the details manually. If you have previous cards associated with your Apple ID, you can add them and enter their CVV number. The other option is to add your card from the bank or card issuer's app.
- After adding the cards, arrange them depending on which you want to be your default card.

Appearance

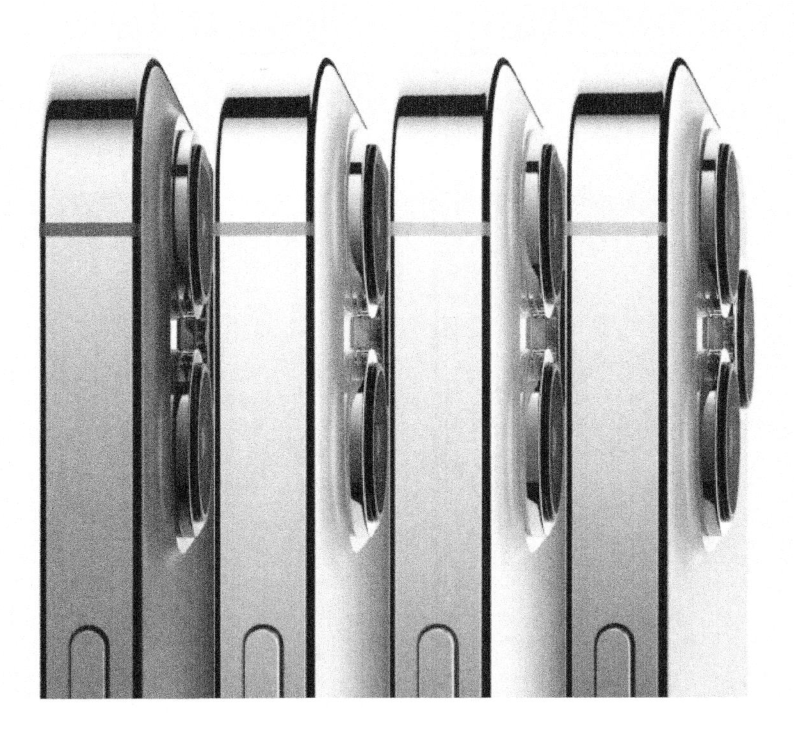

From the design standpoint, the iPhone 13 series has the same design as its predecessor, iPhone 12. Nevertheless, it is slighter thicker and heavier, and includes larger batteries. The width of the iPhone 13 is also 20% less compared to that of the iPhone 12, giving it a sleeker and visually appealing design.

The iPhone 13 models are available in a wide range of colors including PRODUCT RED, green, blue, starlight, and midnight (black). When it comes to the construction, iPhone 13 models feature a flat edge and aerospace-grade aluminum enclosure.

Both iPhone 13 and iPhone 13 mini are crafted with ceramic shield cover glass with nano-ceramic crystals, similar to the previous year's model. This helps to enhance drop protection.

Although there is not much difference from the previous model, you will still need a new case for each iPhone 13 model. This is because some parts of the iPhone like the speaker grill have moved.

Display Zoom

Zoom allows you to zoom in and out certain items in most apps. The Display zoom feature in iPhone allows you to magnify the display and other parts of the screen. The Display Zoom feature works on Apple devices that operate the Latest iOS including iPhone 13 Pro Max.

Follow these steps to enable this feature:

- Open the **Settings** app from the Home screen.
- Navigate to **Display & Brightness** and click to open.

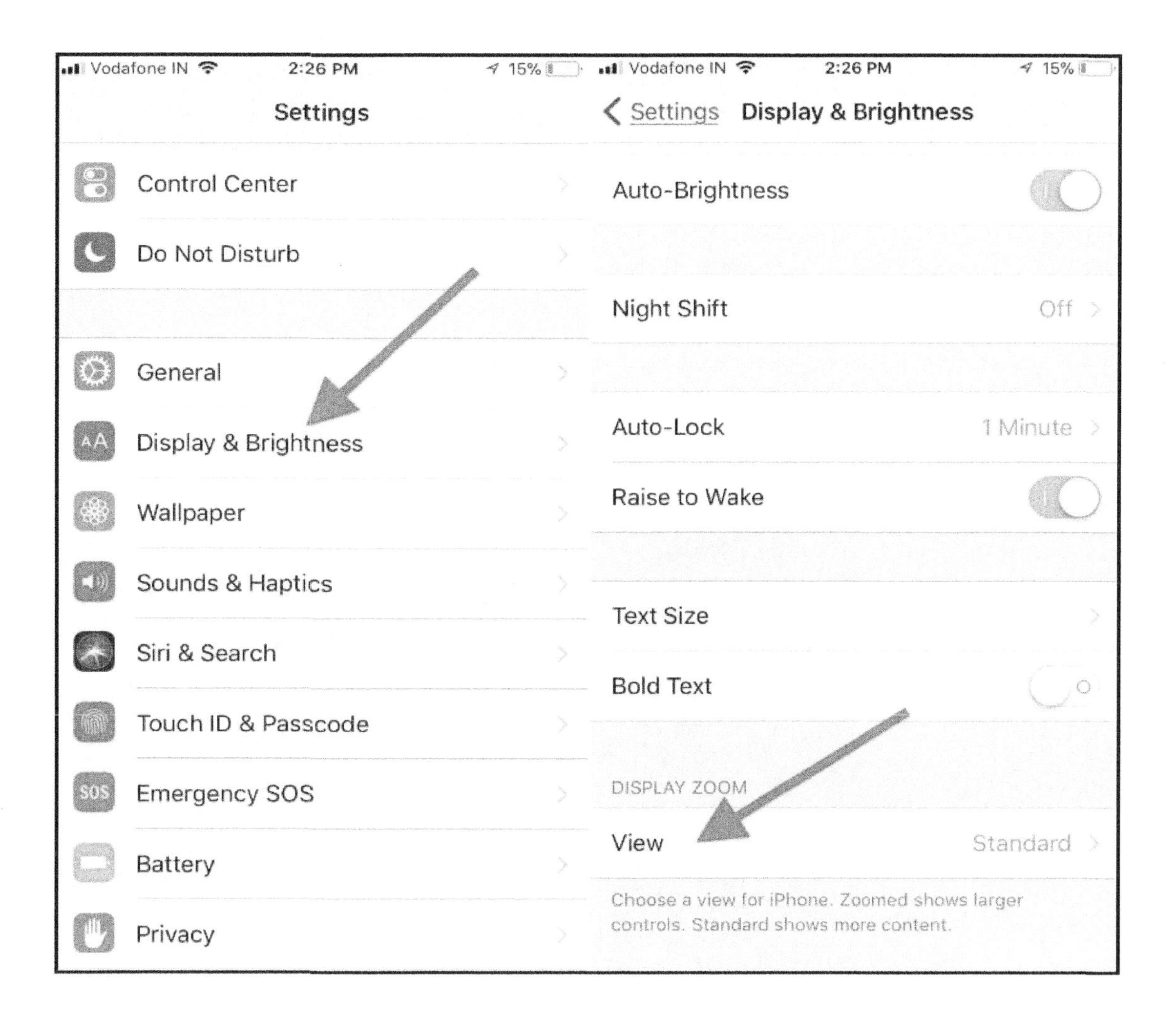

- At the bottom of the screen, scroll down and click on **View**. This is under Display View.
- Click on **Zoomed** and then tap **Set**.

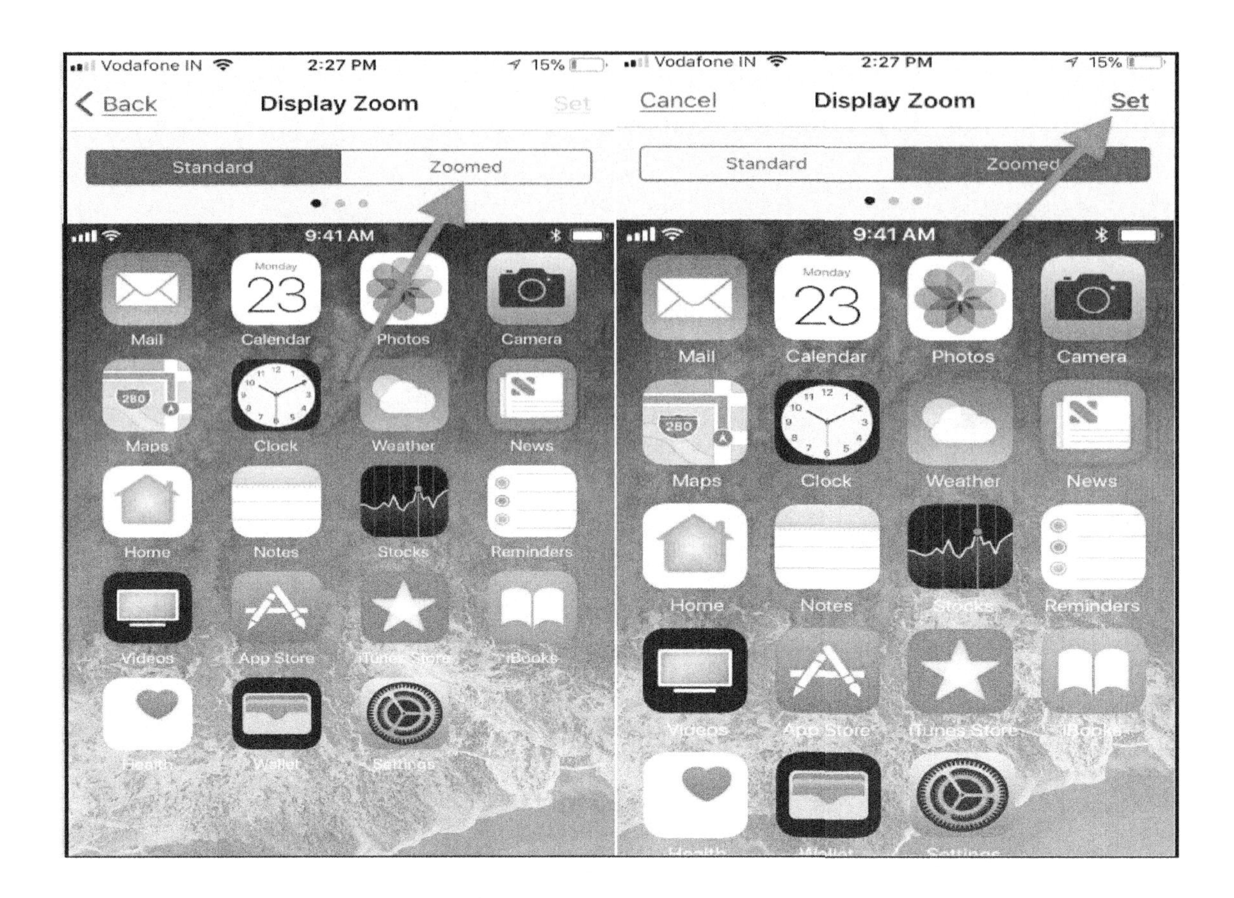

- The option for **Use Zoomed** will pop up on the screen. Click on it.

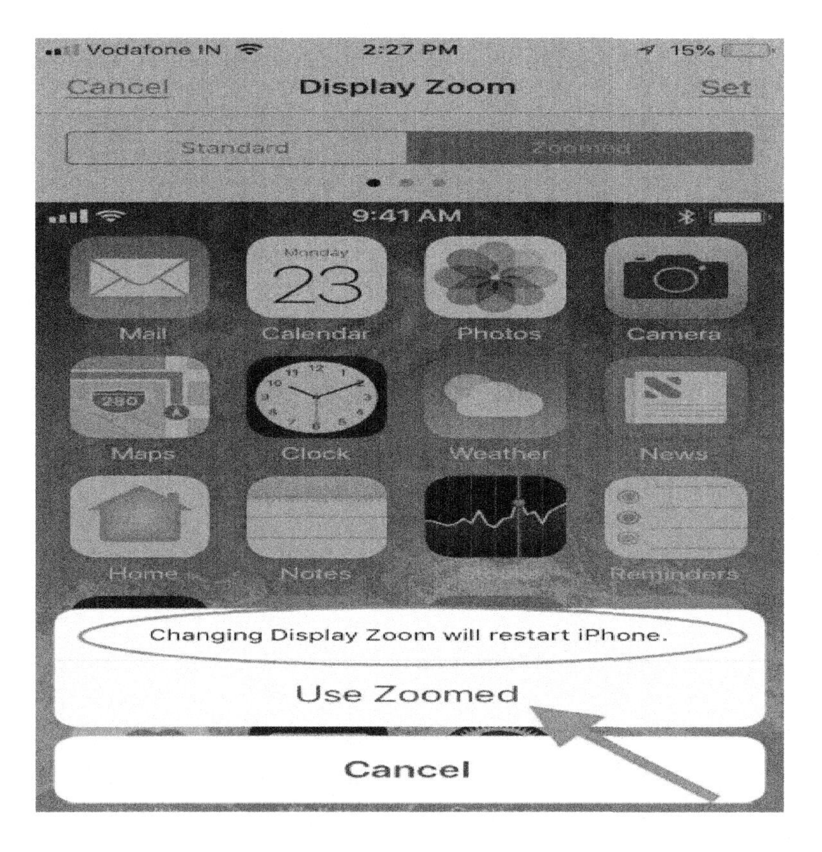

Create a New Apple ID

You can create your Apple ID when you first set up your account, although you can still set it up later in the App Store.

Follow these steps to set your Apple ID in your iPhone 13.

- Open the **Settings** app.

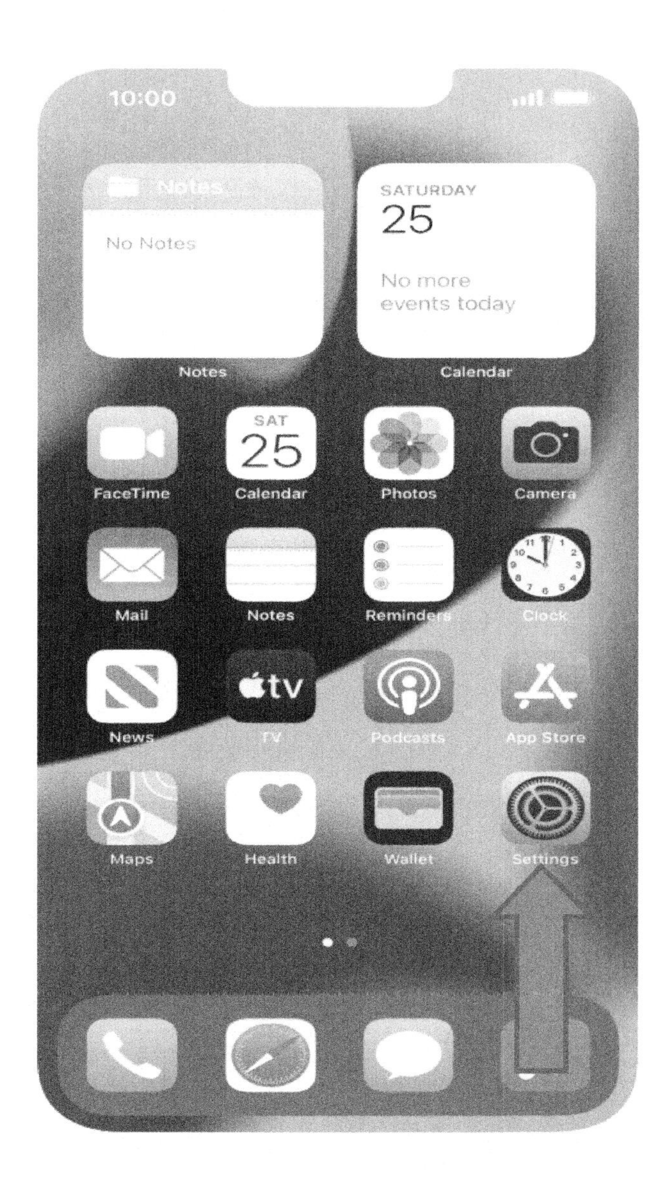

- At the top of the screen, you will see the option to **Sign in to your iPhone**. Click on it.

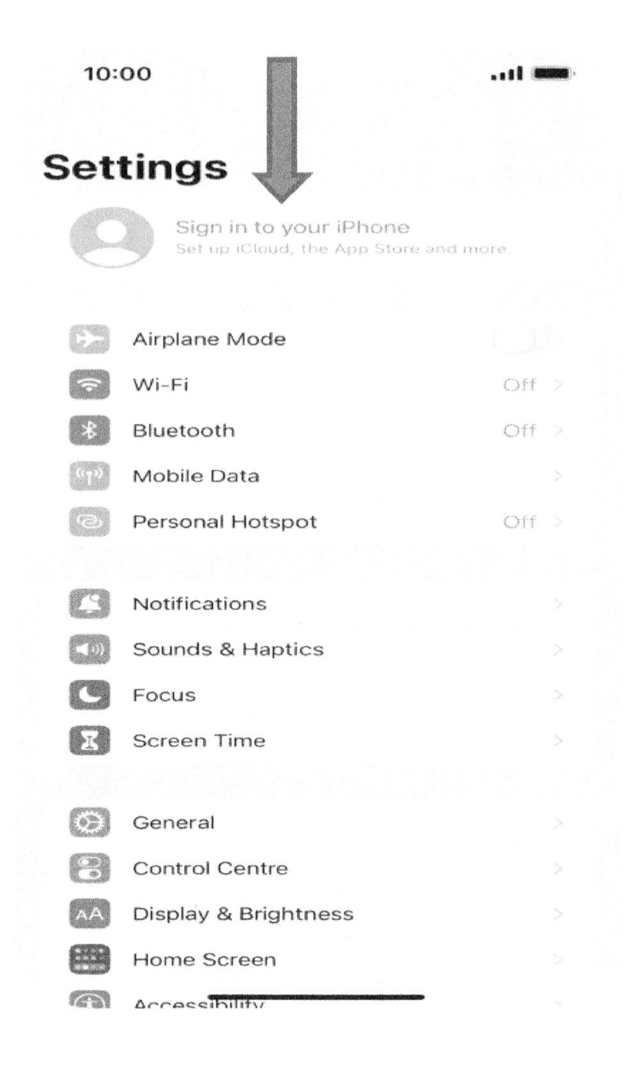

- Tap on the option that you don't have your ID or that you forgot it.
- A window will pop up asking you to **Create Apple ID**. Click on it.
- In the setup instructions, you will be asked to choose your birthday and input your name. Once done, click Next.
- You will also need to provide your email address, which will serve as your new Apple ID. Alternatively, you can tap the option "Don't have an email address" to get a new iCloud email address.
- Toggle on/off the **Apple News & Announcements** button and click **Next**.
- Enter a new password and verify it.
- Next, enter your phone number and choose whether you want to receive a phone call or message for verifying your identity and for recovery of your account if need be.
- Tap next and input the verification code sent to your number.
- Tap **Next** and then **Agree** to the Terms and Conditions.

- Click Agree again to confirm.
- Choose **Merge** or **Don't Merge** when prompted to sync iCloud data from your contacts, reminders, Safari, and calendars.
- You may be prompted to confirm that Find My iPhone is switched on. Click Ok to confirm.

Notification Preferences

When using an iPhone 13, you can enable or disable notifications and change how they are displayed on your lock screen based on your preference. You can even change the notifications for individual apps. Here is how to adjust your phone's notifications to your preference.

- Launch **Settings.**
- Click on **Notifications** and choose the app you want to tweak.
- At the top is the Show Preview option. Here, you need to choose how you want the notification previews to appear from the options provided including **Always**, **Never**, or **When Locked**.

To manage notifications for individual apps:

- Choose the app and turn on or off Allow Notifications.
- Go to Alerts and toggle the type of alerts you want the app to send you. The options include **None**, **Banners**, and **Alerts**.
- The next step is to edit the sound notifications from the app. Click on **Sound** and choose your preferred sound from the options provided.
- Toggle on or off the **Badge switch**.

10:50
Settings
Search
iPhone Tricks.org
Software Update Available
Airplane Mode
Wi-Fi
Bluetooth
On
Cellular
Notifications
Sounds & Haptics
Focus
Screen Time
General

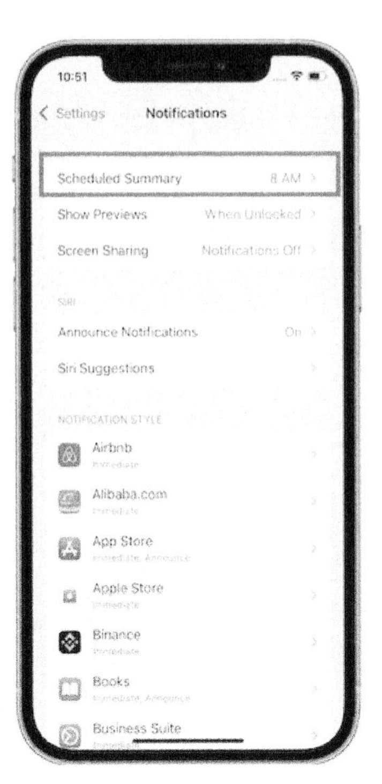
10:51
Settings
Notifications
Scheduled Summary
8 AM
Show Previews
When Unlocked
Screen Sharing
Notifications Off
Announce Notifications
On
Siri Suggestions
NOTIFICATION STYLE
Airbnb
Alibaba.com
App Store
Apple Store
Binance
Books
Business Suite

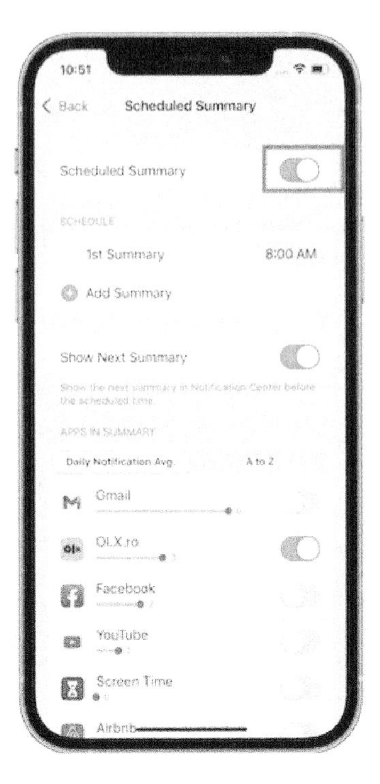
10:51
Back
Scheduled Summary
Scheduled Summary
1st Summary
8:00 AM
Add Summary
Show Next Summary
APPS IN SUMMARY
Daily Notification Avg.
A to Z
Gmail
OLX.ro
Facebook
YouTube
Screen Time
Airbnb

CHAPTER 4: THE BASICS

The Lock Screen

Besides preventing unauthorized access on your device, the Lock screen feature on iPhone 13 also offers a quick and convenient way to access helpful features and information at a glance, even when your device is locked.

When you turn on your iPhone, you can see the current time and date as well your most recent notifications on your lock screen. You can also access the Camera and Control Center without needing to unlock your phone.

To access any of the mentioned features, do the following:

- If you want to open camera, **swipe left.** You can also **touch and hold** the 📷 icon.
- **Swiping down from the top-right corner** allows you to access the control center from your lock screen.
- To scroll through your notifications, **swipe up from the center** of your screen.
- You can **swipe right** if you want to view any of your phone's widgets.

Tip: *You can also control which features you can access from your lock screen by going to Settings and selecting your options.*

How to Lock iPhone 13's Touch Screen

iPhone 13 is set to automatically lock after one minute of no activity. However, you can always change this duration based on your needs.

- Open the **Settings** app and tap **Display & Brightness**.
- Click on **Auto Lock** and choose your preferred time.

To bring your device to sleep mode, press the side button on your iPhone 13 and this will immediately lock your phone. You can then use the side button or tap the screen to wake up your device.

The Home Screen

The iPhone 13 Home screen is more customizable than that of previous models. It is very easy to move apps on this Home screen and you can also reset its layout. Resetting your device's Home screen layout will take it back to its default look and configuration without deleting any settings, apps, or files.

- Locate and open the **Settings** app on the Home screen.
- On the Settings menu, locate the General option and click on it.
- On the bottom of the screen, locate and tap on **Transfer or Reset iPhone**.

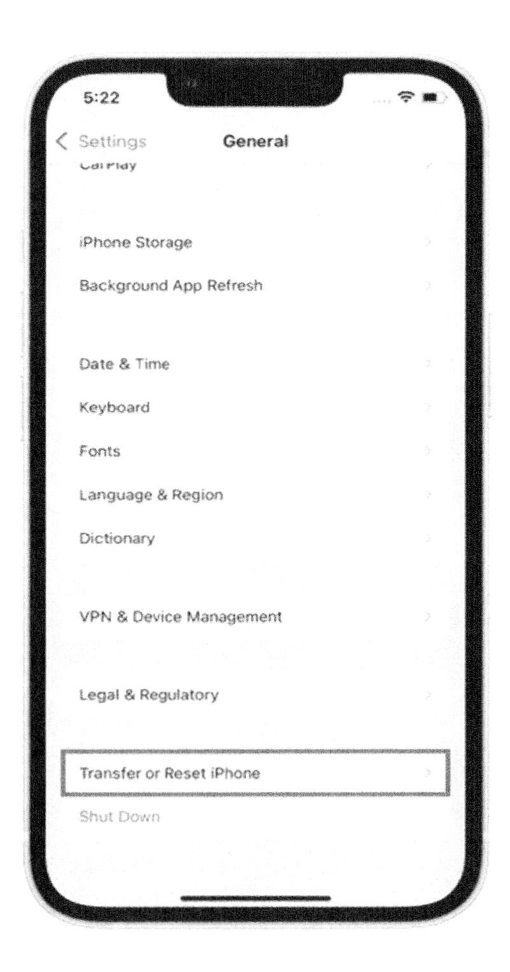

- Information on how to prepare for the new iPhone will pop up on the screen. Click the **Reset** option at the bottom of the screen.

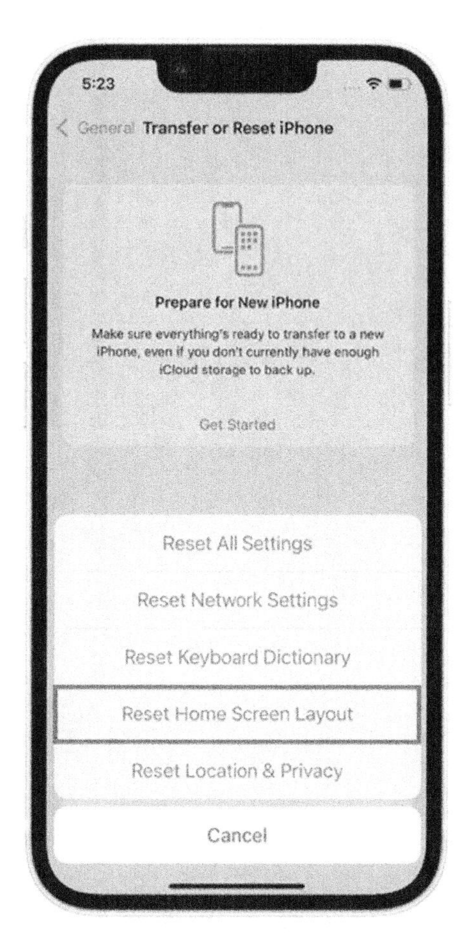

- You will be presented with a list of all the reset options. Select the **Reset Home Screen Layout** option.
- Click **Reset Home Screen** at the bottom of the screen. Doing this will bring your Home screen layout to its factory default settings.

You can also add more widgets to your home screen for more personalization.

To open apps on the Home screen:

- Swap up your phone from the bottom edge of the screen to launch the Home screen.
- Slide right or left to access the apps available on other Home screen pages.
- Tap the app's icon on the Home screen to open it.

You can also choose to customize your lock screen and Home screen's wallpaper. The wallpaper choices can range from dynamic to still images.

To do this:

- Go to Settings then select Wallpaper, then tap Choose a New Wallpaper.
- You can either:
  - Select a preset image from a group found at the top of your screen. Keep in mind that wallpapers marked with will change appearances when you switch on Dark Mode.

- Tap Album and choose from your own photos. You can reposition your chosen image by pinching open to zoom in and then moving the image by dragging it with your finger into your preferred position. To zoom back out, pinch the screen close.

  Note: *You can tap the* icon to toggle the Perspective Zoom option on, which is only available for select wallpaper choices and make it seem like your wallpaper 'moves' every time you change your viewing angle. This option only appears once your Reduce Motion feature is turned off in Settings.

- Tap Set and choose to set it either as a Lock Screen, Home Screen, or both.

If you want to turn on Perspective Zoom for a wallpaper that you've already set, tap the image of the Home Screen or Lock Screen in Wallpaper Settings, and select Perspective Zoom.

How to Use Control Center

iPhone's Control Center gives iOS users instant access to helpful tools like apps, brightness, alarm, dark modem, camera, timer, voice memo, and more. With this feature, you can set your phone on Airplane mode, turn on Wi-Fi, and operate your Apple TV, among many others.

To use the Control Center on your iPhone 13, swipe down your phone's screen starting from the top right-hand corner. Swapping up from the bottom of your phone's screen will close the Control Center.

How to Customize Control Center

You can customize your iPhone's control center depending on which control you intend to use more than the other.

- Open the **Settings** app on your device.

- Scroll down to the **Control Center** option and tap on it.

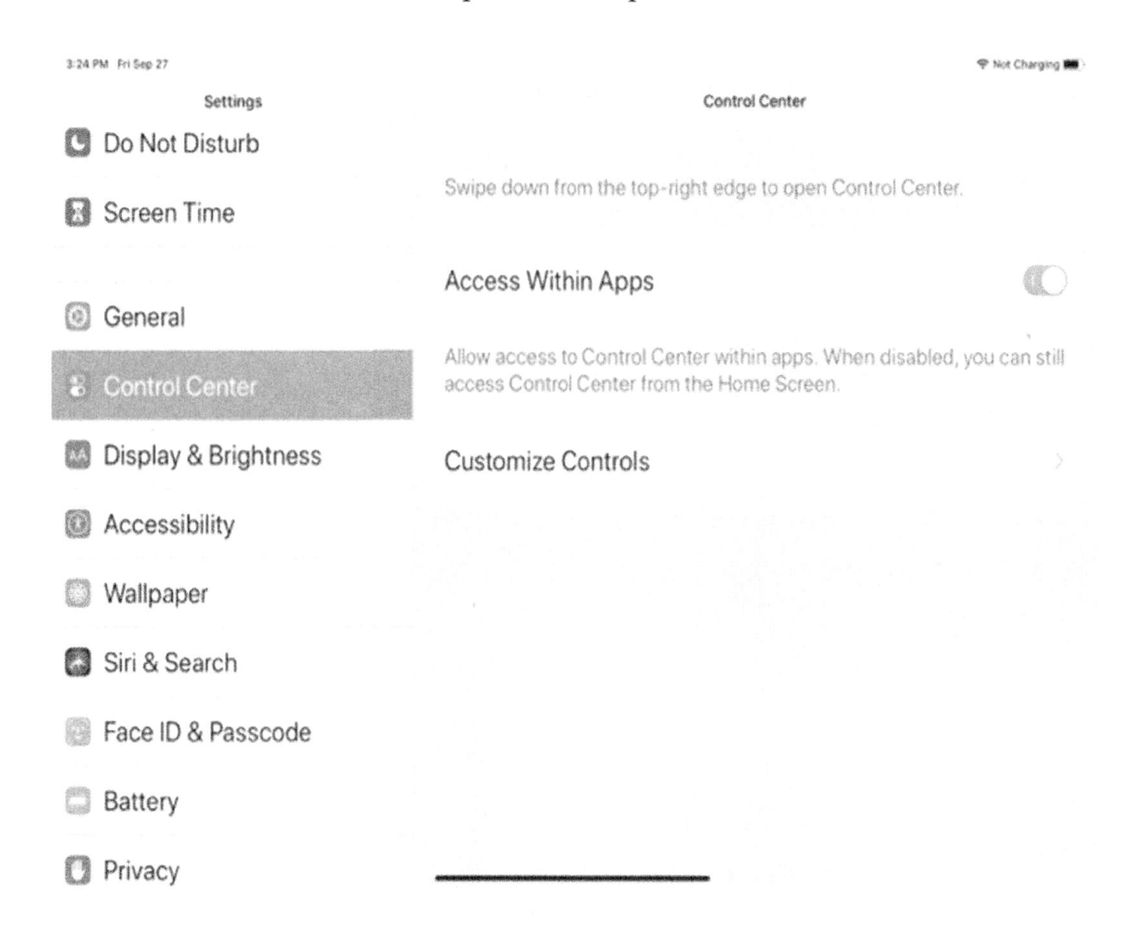

- Tap **Customize Controls**.
- Here, you will have the option to add any optional controls to the Control Center.
- Tap the **+ sign** to add a control and the **– sign** to remove a control.
- If you want to rearrange the controls, click the three lines on the far right of the control.
- Changing the order of the controls is also as simple as dragging the control's hamburger icon to move it up or down the list.

A few features you can do in the Control Center includes:

- Opening the AirDrop options by touching and holding the group of controls found in the top-left and then tapping .
- Touching and holding the icon to take photos or record videos.
- Tapping to disconnect from the WiFi and tapping it again to reconnect. You can also touch and hold the icon to view the name of the WiFi network.
- Disconnecting from Bluetooth devices by selecting the icon. To reconnect, tap the icon again.

Note: *You can switch off access to Control Center in apps by going to **Settings** > **Control Center** then selecting **Access Within Apps** to switch it off.*

The Keyboard

If you are having difficulty using your current keyboard, you can change to another keyboard.

- Launch Settings
- Go to General and then to Keyboard
- You will see the option to Add New Keyboard. Tap on it to see a list of available keyboards.

If you want to add another keyboard while typing, hold and press the globe or smiley face icon. Then click another keyboard. However, if you want to remove a keyboard, go to **Edit** and click on the minus sign.

A unique spec for the iPhone keyboard is the Haptic feedback. This is a clicking sound that occurs each time you press a key on your keyboard. This touch-based response makes your iPhone vibrate, making it a fun way to interact with your iPhone.

Unfortunately, this feature is only built-in to Android devices and iOS users need to install a third-party app to enjoy this feature. A good choice is Google's Gboard.

- On your iPhone, download the Gboard iOS app.
- Go to your device **Settings** and open it.
- Click on the **General** option and then select Keyboard.
- On the upper part, tap **Keyboards**.
- You will see the Add New Keyboard option. Click on it.
- Click on **Gboard** and then select **Gboard** again.
- Turn on the **Allow Full Access** option.
- Launch the Gboard app and navigate to the **Keyboard Settings** option.

- Locate the Enable haptic feedback on the press option and tap it.

Once the haptic feedback is enabled, your keyboard will vibrate when you open apps that use it. If you want to turn off this feature, go to **Accessibility** and then **Touch** and click on **Vibration**.

How To Cut, Copy, And Paste

What To Know

1. To Copy Text: Tap and hold until the first word is highlighted. Drag until you've highlighted all the text you want to copy, then tap Copy.

2. To Copy A Link: Tap and hold the link, then tap Copy from the menu. Tap and hold the image, then tap Copy to copy an image.

3. To Paste: In the app where you want to paste what you copied, double-tap or tap and hold, depending on the app, then select Paste.

How To Copy And Paste Text On iPhone

- The copy and paste commands are accessed through a pop-up menu. Most apps support this feature.
- Find the text you want to copy.
- Tap a word or area of the screen and hold your finger down until a window appears that magnifies the text you selected. When it shows up, remove your finger. The copy and paste menu appears, highlighting the word or section of text you tapped.
- Drag the handles (circles) at the edges of the highlighted word or section to select more text. Tap and drag either of the blue lines in the direction you want to select,left and right, up and down. The menu reappears when you.
- When the text you want to copy is highlighted, tap Copy. The copied text saves to a virtual clipboard. After you tap an option on the menu, the menu disappears.
- Go to the app you want to copy the text into. It can be the same app you copied it from, like copying text from one email to another in Mail or another app, such as copying something from the Safari web browser into a to-do list app.

- Tap the location in the app or document where you want to paste the text and hold your finger down until the magnifying glass appears. When it does, remove your finger, and the pop-up menu appears. Tap Paste to paste the text.
- When you paste into a document that contains words, drag your finger after the magnifier appears to place the cursor where you want the new text to appear.

How To Copy Links On An iPhone

To copy a link, tap and hold the link until a menu appears from the bottom of the screen with the URL of the link at the top. Tap Copy. Paste it using the same steps as other text.

How To Copy Images On An iPhone

You can also copy and paste images on the iPhone. To do that, tap and hold the image until a menu pops up from the bottom with Copy as an option. Depending on the app, that menu may appear from the bottom of the screen.

- **Advanced Features:** Look Up, Share, and Universal Clipboard

The copy-and-paste menu contains more options than those. Here are some of the highlights:

- **Look Up:** To get the definition for a word, tap and hold the word until it's selected. Then, tap Look Up to get a dictionary definition, suggested websites, and more.
- **Share:** After you copy text, pasting isn't the only thing you can do. You might prefer to share it with another app such as Twitter, Facebook, or Evernote.
- **Universal Clipboard:** If you have an iPhone and a Mac, and both are configured to use the Handoff feature, use the Universal Clipboard to copy the text on your iPhone and paste it on your Mac, or vice versa, using iCloud.

Add a Widget

If you are looking for an easy way to edit your iPhone 13's home screen, consider adding a widget. Adding widgets to your home screen allows you to personalize your iPhones so you can get access to relevant information and tasks from your favorite app.

You can get information on the weather, upcoming events, top news stories, favorite photo albums, and more right from your lock screen without necessarily opening the app.

But how do you include widgets on your iPhone 13? Here are some steps to follow:

- Press and hold an empty area on your home screen until the **+ icon** appears on the top left side of your device.
- Tap on the plus **(+)** icon to open the widgets menu.
- You will see a list of widgets that are available for your use. Navigate through the list until you find the widgets from your installed apps that you would like to add to your home screen. If your preferred widget is missing from the list, click on the search field and follow the on-screen instructions to find the widget.
- Once you have the widget that you would love to add to your iPhone device, the next step is to choose the widget size. To do this, swipe your finger right or left until you find your preferred widget size.
- Next, click on the **Add Widget** option and your widget will automatically appear on the home screen.
- To finish the process of adding a widget to your device, tap the **Done** option on your screen's top-right corner.
- Tap and hold the widget to drag it to the place of your choice on your home screen. If you want to access the Today View, slide your home screen right from the left edge. Then scroll up and down and the widgets will appear in Today View.

Quick Tip: You can create a Smart Stack that contains several widgets displaying relevant information like time, location, and useful activities. This is an easy and quick way to get all the information you need.

Airdrop

AirDrop is a useful iPhone feature that lets you share content wirelessly from one device to another compatible device. With this neat little feature, iOS users can effortlessly send and receive contacts, photos, videos, map locations, and much more between Apple devices via Bluetooth technology.

How to Activate AirDrop

Follow these steps to turn on AirDrop and make it visible.

- Launch the **Setting** option on your device.

- Navigate to the **General** option and click on it.
- Scroll down to **AirDrop** and tap on it.
- You will be provided with three options. Choose **Everyone** to make your iPhone device discoverable.

How to Share Content Using AirDrop

- Go to the file you want to share and choose the Share icon.
- Click on AirDrop and find the nearby iOS device or AirDrop user.
- Click done to begin the sharing process.

Receiving a file through AirDrop is easy and you just need to **Accept** or **Decline** the incoming file.

Is Airdrop Not Working? Try These Tips

- AirDrop is brilliant - when it works - but it can be a little glitchy. Here are a few tips to try if AirDrop isn't working for you on your iPhone, iPad, iPod Touch, or Mac.
- Check both devices sending and receiving have Wi-Fi and Bluetooth switched on.
- Make sure any Personal Hotspots are turned off.
- Make sure both devices are within 9 meters (30 feet) of each other
- If set to receive from Contacts Only, who must sign both devices into iCloud, and the email address or phone number associated with the sender's Apple ID must be in the Contacts app of the receiving device.
- If you see Receiving Off on your iPhone, iPad, or iPod Touch and can't tap to change it: Go to Settings > Screen Time > Content & Privacy Restrictions > Allowed Apps > Make sure that AirDrop is turned on.
- On Mac, Choose Apple menu > System Preferences > Security & Privacy > Click the Firewall tab > Click the lock > Enter your password when asked > Click Firewall Options > Deselect "Block all incoming connections".

Before You Begin

- Ensure that the person you're sending to is nearby and within Bluetooth and Wi-Fi range.
- Check that you and the person you're sending have Wi-Fi and Bluetooth turned on. If either of you has Personal Hotspot on, turn it off.
- Check if the person you're sending to has their AirDrop set to receive from Contacts Only. If they do, and you're in their Contacts, they need to have your Apple ID's email address or mobile number in your contact card for AirDrop to work.
- If you're not in their Contacts, have them set their AirDrop receiving setting to Everyone to receive the file.
- You can put your AirDrop receiving setting to Contacts Only or Receiving Off at any time to control who can see your device and send you content in AirDrop.

How To Use Airdrop

Follow the steps for your device:

On your iPhone X, 11 or later

- Open an app, then tap Share or the Share button. If you share a photo from the Photos app, you can swipe left or right and select multiple photos.
- Tap the AirDrop button.
- If the person you want to share with also has an iPhone 11 or later model, point your iPhone in the direction of the other iPhone.
- Tap the profile picture of its user at the top of the screen. Or you can use AirDrop between your own Apple devices. If you see a red numbered badge on the AirDrop button, there are multiple devices nearby that you can share with. Tap the AirDrop button, then tap the user you want to share with. Learn what to do if you don't see the AirDrop user or your other device.

How To Use Airdrop On Mac

- When you are ready to share a file using AirDrop on your Mac, open the file you want to send and click the Share button in the app window. You can also Control-click the file in the Finder and choose Share from the shortcut menu.

- You'll then need to choose AirDrop from the sharing options and select the AirDrop recipient from the AirDrop box. You can also launch the Finder and open the AirDrop from the left-hand bar. It's then possible to drag and drop the file onto the recipient.

- You'll need to tap Accept or Reject when the request appears as a notification or as a message in the AirDrop window for receiving files. If you accept, the file will save into your Downloads folder.

How To Accept Airdrop

- When someone shares something with you using AirDrop, you see an alert with a preview. You can tap Accept or Decline.

- If you tap Accept, the AirDrop will come through within the same app it was sent from. For example, photos appear in the Photos app and websites open in Safari. App links open in the App Store to download or purchase the app.

- If you AirDrop something to yourself, like a photo from your iPhone to your Mac, you won't see an option to Accept or Decline, it'll automatically get sent to your device. Just make sure that both devices are signed in with the same Apple ID.

How To Adjust Airdrop Settings

To choose who can see your device and send you content in AirDrop:

- Go to Settings, tap General.

- Tap AirDrop, then choose an option.

You can also set your AirDrop options in Control Center. Here's how:

- On iPhone X or later, swipe down from the screen's upper-right corner to open Control Center. Or follow the same motion to open Control Center on your iPad with iOS 12 or later or iPad. On your iPhone 8 or earlier or iPod touch, swipe up from the bottom of the screen.
- Press firmly or touch and hold the network settings card in the upper-left corner.

Touch and hold the AirDrop button, then choose one of these options:

- **Receiving Off:** You won't receive AirDrop requests.
- **Contacts Only:** Only your contacts can see your device.
- **Everyone:** All nearby Apple devices using AirDrop can see your device.

If you see Receiving Off and can't tap to change it:

- Go to Settings > Screen Time.
- Tap Content & Privacy Restrictions.
- Tap Allowed Apps and make sure that AirDrop is turned on.

Allow Others To Send Items To Your Iphone Using Airdrop

- Open Control Center, touch and hold the top-left controls, then tap.
- Tap Contacts Only or Everyone to choose who you want to receive items from.
- You can accept or decline each request as it arrives.

How to Dictate Text

If you don't want to type text on your device, Apple provides the option to dictate the text instead. The general text dictated can also be processed in many languages without requiring any internet connection. You can also dictate text of any length without being timed out. You can choose to stop dictation manually. Otherwise, it stops automatically when 30 seconds have passed without you speaking.

- But first, you need to enable dictation on your iPhone.

- Launch the **Settings** apps and tap the **General** option.
- Click on Keyboard to launch it.
- Locate the **Enable Dictation** option and turn on the switch next to it.

Important: *The dictation feature may not be accessible in all languages, as well as all countries or regions. Cellular data charges may apply.*

How to Use Dictation on iPhone

Now that you have enabled dictation, follow these steps to use it.

- Open any app such as messages that use the keyboard.
- On the onscreen keyboard, tap on the microphone, usually located between the spacebar and the Emoji button.
- Speak and then tap **Done**.

Tip: *You can insert dictated text by tapping the place you want to insert the text into and then speaking. This is the same for replacing text.*

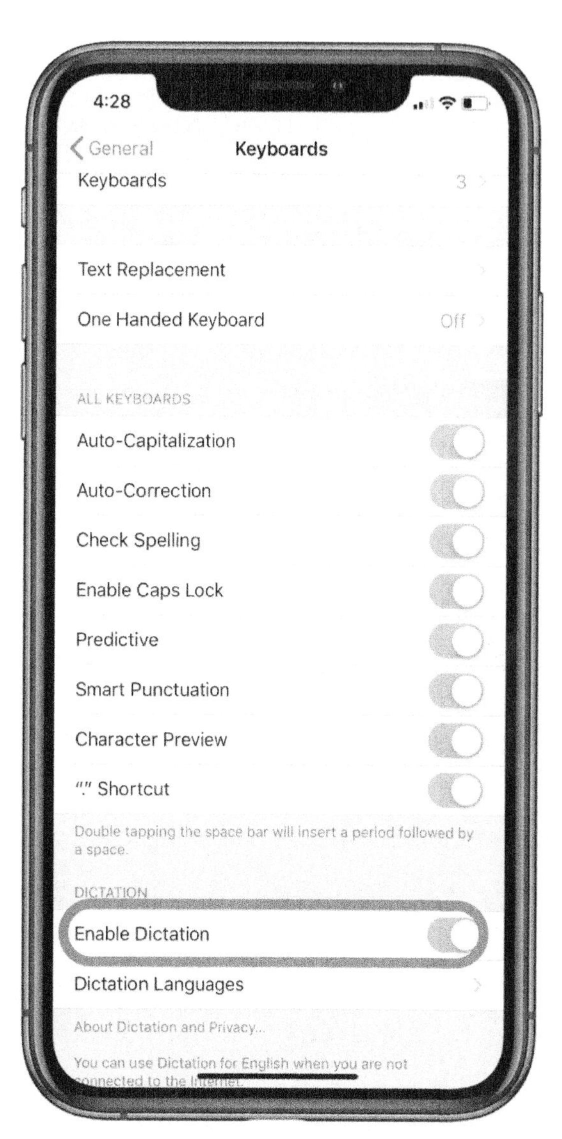

If you want to add punctuation or formatting to your dictated text, you can include the following commands:

- Period
- Question Mark
- Dollar Sign
- Comma
- New Line
- Exclamation Mark
- Open Parenthesis... Close Parenthesis
- Cap – to capitalize a word
- Colon
- Semicolon

- New Paragraph
- Quote.... End Quote
- Caps On...Caps Off – if you want to capitalize the first letters of the enclosed words
- Hashtag
- Winky – to insert the ;-) face
- Smiley – to insert the :-) face
- Frowny – to insert the :-(face
- No Space On... No Space Off – if you want to a series of words together ***(may not be available for all languages)***
- No Space – if you want to remove the space between words ***(may not be available for all languages)***
- All Caps – to write the next word in all uppercase letters
- No Caps On... No Caps Off – if you want the enclosed words to be written in all lowercase letters
- All Caps On... All Caps Off – if you want the enclosed words to be written in all uppercase letters

By saying any of the listed commands, your text "Hello exclamation mark cap how are you question mark" will show up as "Hello! How are you?" on screen.

How to Cut, Copy, and Paste

iPhone has a universal clipboard that lets you cut, copy, and paste content from your iPhone device to another iOS device. Here are some tricks for selecting the text and copying on iPhone 13, and paste it elsewhere.

Step 1: Select Text

- Double-tap to select a word
- Triple tap with one finger to highlight sentences
- Quadruple tap to select a paragraph

The highlighted text should now be yellow.

Step 2: After selecting the text, the next step is to edit it. Here are some editing options:

- To cut, pinch the screen closed with three fingers. Alternatively, tap the **Cut** option on the banner right above the text.
- To copy, pinch the screen closed with three fingers twice. You can also tap the **Copy** option.
- To paste, pinch the screen open with three fingers or tap **Paste**.

Step 3: To insert or edit the text, touch and hold the text to magnify it. Then move the insertion point where you want to insert it by dragging it.

Step 4: Click on the paste option or pinch out with three fingers to paste>

Passwords

It is easy to find passwords saved on your iPhone as Siri can help you with this. Alternatively, you can also locate your saved passwords through the Settings app.

- Open the Settings app and click on **Passwords.**
- You will be required to enter your passcode or you can use your Face ID or Touch ID to log in.
- This will give you access to the websites with saved passwords. Click on the website of your choice to reveal the password.

You can also choose to set iPhone into erasing all information, media and personal settings after 10 failed consecutive password attempts.

To enable this:

  - Go to **Settings** and select **Face ID & Passcode.**
  - Switch on the **Erase Data** option.

When all data has been erased, you can restore your device from a backup or set it up as a new phone.

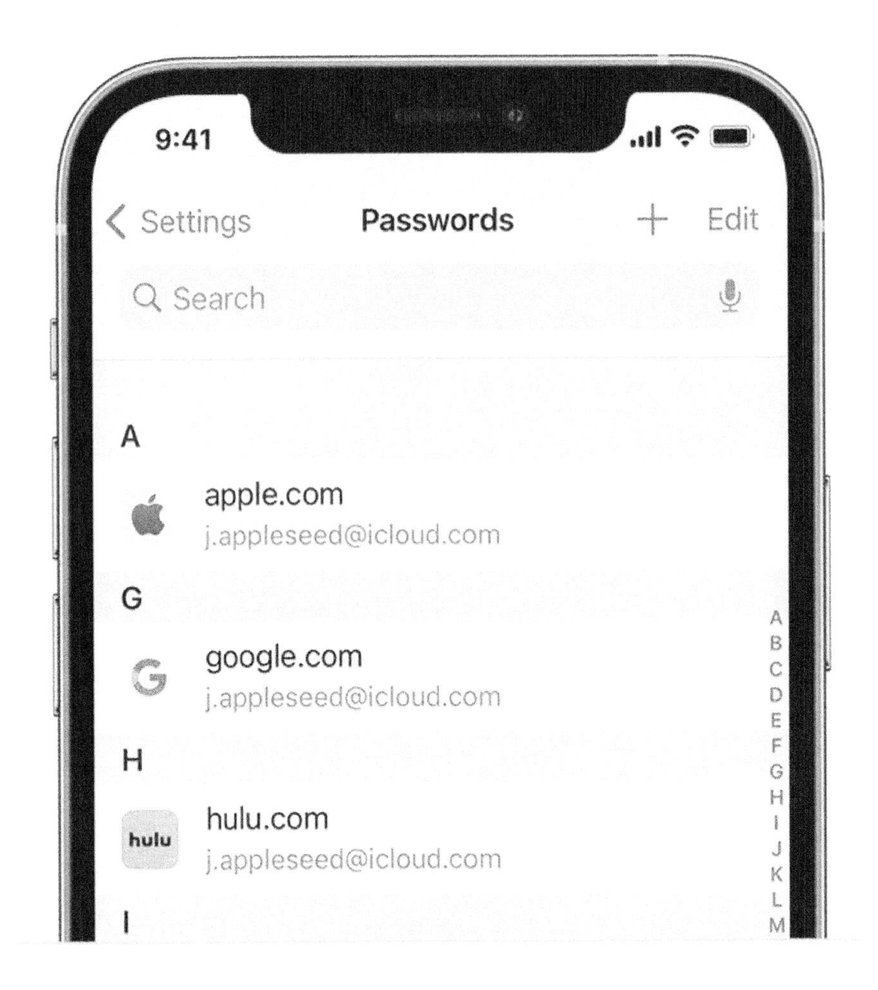

- To update the password, select the **Edit** button on the upper right side of the screen.
- On the bottom part, you will see the option to **Delete Password** if you want to.

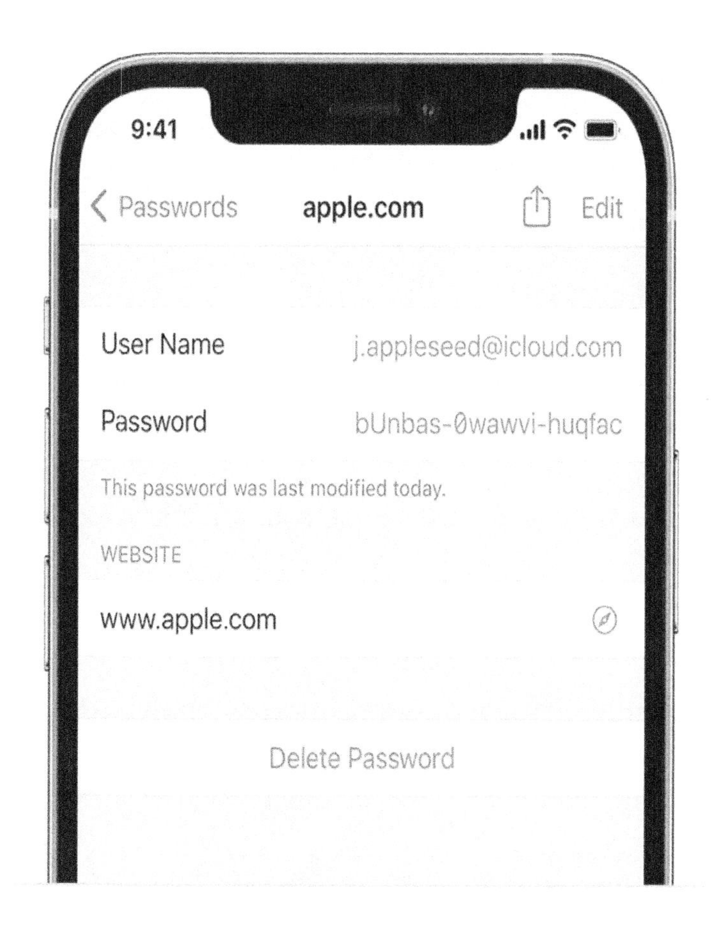

For a more convenient way of unlocking your phone, you can set up Face ID by:

- Going to **Settings > Face ID & Passcode > Set up Face ID.** Then, follow the instructions that pop up on your screen.
- If you want to set up an alternate appearance for Face Recognition, go to **Settings > Face ID & Passcode** then tap **Set Up an Alternate Appearance.** Then, follow the instructions that show up on screen.

For those with physical limitations, you can adjust how Face ID recognizes you by tapping Accessibility Options in the Face ID Set Up interface. Doing this means that facial recognition will not be requiring the full range of head motion, but it does require more consistency when looking at your phone.

If you're blind or have low vision, there is an Accessibility Feature that is also available in Face ID. Go to **Settings > Accessibility** then turn off the **Require Attention for Face ID** option. This way, you won't have to look at your iPhone with your eyes open to unlock it.

Face ID also works while you're wearing a mask and/or glasses. To enable this:

- Go to **Settings > Face ID & Passcode.**
- Select the **Face ID with a Mask** option to turn it on, then follow the instructions that show up on screen.
- If you want to add glasses to your appearance, select **Add Glasses** then follow the instructions that show up on screen.

Tip: *Wear a pair of transparent glasses instead of sunglasses if you want to improve the accuracy of the facial recognition feature.*

You can also choose to temporarily prevent Face ID from unlocking your phone. To do this:

- Press and hold the side button and the volume button of your iPhone simultaneously for two seconds.
- When the slider appears, pressing the side button will immediately lock your iPhone.

Face ID will be enabled again after you use your passcode to unlock your iPhone.

How To Find Your Accounts And Passwords On iPhone Or iPad

- Open Settings on your iPhone or iPad.
- Tap Passwords & Accounts.
- Tap Website & App Passwords. Authenticate as needed with Face ID or Touch ID.
- Tap the search field to search for an entry if you need to.
- Tap the entry you're looking for.
- Tap on a username/email address or password to copy one of them.
- Tap Copy to copy the username/email address or password.

How To Add Accounts And Passwords On iPhone Or iPad

iOS has a built-in mechanism for adding accounts and passwords to your iCloud Keychain. When you start to create an account, Safari will suggest a password for you. If you choose to use that password, Safari will also save your account details to iCloud Keychain. If you'd like to add accounts and passwords on iOS manually, here's how!

- Open Settings on your iPhone or iPad.
- Tap Passwords & Accounts.
- Tap Website & App Passwords. Authenticate as needed with Face ID or Touch ID.
- Tap the + button in the top-right corner of the screen.
- Type in the Website, Username, and Password fields where you'll use these credentials.
- Tap Done.

How To Delete Accounts And Passwords On iPhone Or iPad

- Open Settings on your iPhone or iPad.
- Tap Passwords & Accounts.
- Tap Website & App Passwords. Authenticate as needed with Face ID or Touch ID.
- Tap Edit in the top right corner.
- Tap to select the credential(s) you want to delete.
- Tap Delete in the upper-left corner.
- Tap Delete.

The password will be removed from iCloud Keychain and will no longer be accessible when you access the website associated with it.

What Is Icloud Keychain?

iCloud Keychain is Apple's native password manager supported across iPhone, iPad, iPod, and Mac devices. It allows you to keep your website and app passwords, along with credit card information, Wi-Fi network information, and other account information, up to date across all Apple devices approved and associated with your Apple ID. It can also keep the accounts you use in Mail, Contacts, Calendar, and Messages up to date. To use Apple iCloud Keychain, an Apple device needs to be running iOS 8.4.1 or later, iPad 13 or later, and macOS X 10.10.5 or later. Here's everything you need to know about Apple's password manager, iCloud Keychain, how it works, and how to set it up to remember your passwords.

How Does Icloud Keychain Work?

Apple's iCloud Keychain is secured with 256-bit AES encryption during storage and transmission. The data collected cannot be read by Apple and is protected by a key made from information unique to your device and combined with your device passcode.

When iCloud Keychain is set up, passwords and credit card information - though not the security code - along with other account information like usernames and Wi-Fi passwords will be automatically filled in.

For some Apple devices, like iPhones, you will need to authenticate yourself using Touch ID or Face ID to fill in the respective information.

How To Set Up iCloud Keychain On iPhone, iPad, And iPod Touch

To set up Apple iCloud Keychain on iPhone, iPad, and iPod Touch, follow these steps:

- Open Settings on your iPhone, iPad, or iPod Touch
- Tap your name at the top
- Tap on iCloud
- Tap on Keychain
- Toggle Keychain on. You might be asked for your Apple ID and password

How To Set Up Icloud Keychain On Mac

To set up iCloud Keychain on Mac, follow these steps:

- Open the Apple Menu in the top left corner of your screen
- Tap on System Preferences
- Click on Apple ID
- Tap on iCloud from the sidebar
- Tick Keychain to turn it on
- Enter your Apple ID and password

You need to be running two-factor authentication on Mac to use iCloud Keychain.

How To See Saved Passwords And Access Icloud Keychain On Iphone

You can ask Siri to find your saved passwords and ask Siri to find a specific password for one particular site. For example, if you say, "Hey Siri, what is my password for Netflix", it should return the result.

Alternatively, you can follow these steps to see and edit your saved passwords on iCloud Keychain on iPhone, iPad, and iPod Touch:

- Open Settings on your iPhone, iPad, or iPod Touch
- Select Passwords
- Or select Passwords and Accounts and then Website and App passwords if running iOS 13 or earlier
- Use Touch ID or Face ID when prompted to view passwords
- Tap on a website to see the password
- Tap Edit in the top right to edit a password
- Tap Delete Password at the bottom to delete it

How To Access Icloud Keychain On Mac

As with iPhone and iPad, you can use Siri to find your passwords on Mac. Alternatively, to find saved passwords on Mac, follow these steps:

- Open Safari
- Choose Preferences from the Safari menu in the top left corner
- Click on Passwords
- Sign in with Touch ID, enter the password for your Mac or authenticate your password using an Apple Watch running watchOS 6 or later
- Select a website to see the password
- Tap on Details when a website is selected to update a password and tap Done after
- Tap on Remove when a website is selected to delete a password

How To Delete Passwords On Icloud Keychain

To delete a password from iCloud Keychain on an iPhone, iPad, or iPod, follow the steps below:

- Open Settings
- Tap on Passwords
- Tap on the website for the password you want to delete
- Tap 'Delete Password' at the bottom

To delete a password from iCloud Keychain on Mac, follow these steps:

- Open Safari

- Tap on Preferences in the Safari menu
- Click on Passwords
- Select the website you want to delete the password from
- Tap Remove at the bottom of the list

How to Share Something

There are several ways to share files, pictures, and other data on iPhone 13. You can choose to send an item using AirDrop or even through Siri. Apple introduced the share button in 2014 and it has become an important feature in all its series.

This button allows you to share items like pictures and documents between apps and with other people. However, there are notable changes in the share sheet in iOS in terms of the design and colors. With this new share sheet, there is a scrollable list of actions and shortcuts.

Show Your iPhone Screen on the TV

Screen mirroring is a quick way to stream and share content from your device to your TV. You can use the regular screen mirroring or use AirPlay to stream any file to a compatible TV.

Follow these steps to screen mirroring your iPhone 13's screen on the TV:

- The first step is to open the Control Center. To do this, swipe down from your screen's top-right corner.
- Click on **Screen Mirroring** and two rectangles will appear.
- This will prompt your device to search for compatible devices. Click on the device of your choice.
- After a little while, you should be prompted to input your passcode that appears on the TV.
- Enter the passcode on your iPhone to start screen mirroring.

To show your iPhone screen on TV through AirPlay:

- Make sure your iPhone is connected to the same WiFi network as your Apple TV or AirPlay compatible TV.
- Select the video that you want mirrored on your TV screen.
- Select the Airplay icon on your iPhone.

Note: Some apps might require you to tap different buttons first before selecting AirPlay. The Photos app, for example, you would need to tap the Share ⍐ button first.

- Then, select your Apple TV or your AirPlay compatible TV.

If you want to stop mirroring your video through AirPlay, go to the app that you're playing the video from then tap the AirPlay option and then select your iPhone device from the list.

Note: *There are some apps that might not support AirPlay mirroring. If this happens, check the App Store on your Apple TV or AirPlay compatible TV and find out if the app is available.*

CHAPTER 5: APPS & APPSTORE

Apps Included in Your iPhone

Like all iPhone devices, the iPhone 13 comes with useful apps that will give you an exceptional user experience. Some of the apps that you can expect in your new device include:

- **Yelp**

This is an impressive app that can help you find new places, book appointments, and get quotes. It also allows you to filter your search to make researching easy.

- **Duolingo**

If you enjoy learning new languages, then you will be pleased to know that iPhone 13 comes with the Duolingo app. This app offers three dozen languages and free basic lessons that you can take at your own convenience.

- **Libby**

The Libby app can give you access to a wide collection of audiobooks and ebooks from libraries worldwide. This free app lets you add bookmarks and notes, change the playback speed for audiobooks, and tweak the text size, among many others.

- **Tweetbox**

Although Twitter is a great social media platform, it can be extremely difficult for some people to get by. This is where Tweetbox comes in handy. Tweetbox can help chronologically organize your tweets, making it easy to access and use the service. It eliminates ads and promoted tweets, which can be at times annoying. However, this app requires a subscription fee of $1 monthly or $6 annually.

- **Lose It**

This is a third-party app that Apple has allowed into this device for iOS users that want to attain their weight loss goals. You simply need to input your target weight and Lose It will calculate the number of calories you need to consume every day to aid in your weight loss efforts.

How to Remove Default Apps

Your iPhone 13 device comes with many default apps and most of these apps are quite useful. However, we cannot ignore that some of the pre-installed apps are nearly never used and only take up valuable space in your phone. Fortunately, it is possible to delete some of these apps. Built-in apps that can be deleted from your device include:

- Calendar
- Home
- Videos
- Files
- Apple Books
- Facetime
- Tips
- Weather
- Music
- Voice Memos
- Podcasts
- Mail
- Notes
- Reminders
- Activity
- Compass
- Calculator
- Watch App
- Stocks
- TV
- Measure
- News
- iTunes Store
- Map

On the other hand, there are in-built apps that you can't remove from your device. They include:

- App Store
- Phone
- Find My
- Camera
- Photos
- Messages
- Settings
- Clock
- Safari

How to Delete Default Apps:

- Long press the app's icon and then tap the **Remove App** command. Alternatively, you can hold your finger on the app icon until it starts jiggling. Then, click the minus button next to the icon to delete the app.
- You will then be asked to confirm whether you want to delete the app. Confirm to remove the app from your iPhone.

NB: It is worth noting that deleting any of the in-built apps from your iPhone 13 will remove any related configuration and user data. An exception to this is any content that you created on the cloud.

Explore the AppStore

The Apple App Store is dedicated to helping iPhone users find and shop the latest products and accessories from Apple. Here, you can get to explore new apps, in-app events, and games.

To buy an app from the App Store

- Tap Get if the app is free. If it comes at a cost, tap the price.
- You will then be required to authenticate your Apple ID by entering your passcode or with your Face ID or Touch ID.
- Once the purchase is complete, the app will download and be added to the App Library.

You can see new app releases, browse by category, or see the top charts. To learn more about an app, be sure to check the:

- Ratings and reviews
- In-app events
- Supported languages
- File size
- Previews
- Screenshots
- Compatibility with other Apple devices

To Share or Give an App

If you want to share your app with someone else:

- Select the app to view its details.
- Tap the icon and select a sharing option. You can also choose the **Gift App** option, although it may not be available for all applications.

Top Social Media Apps

1.TikTok

It is not surprising that TikTok is among the top social media apps on iPhone 13. This app gives users access to countless videos based on different topics. It is very easy and intuitive to use compared to other social media apps and is a great way to enhance interactivity.

With this app on your phone, you can create short and informative or amusing videos and share them with others. You will also access content from other TikTok users. It is undoubtedly the top destination for any mobile video.

You can also take your original videos to the next level by adding special effects, music, filters, and other fun stuff. With TikTok, you can shoot as many times as needed with the pause and resume features.

It also has millions of free music clips and sounds you can integrate into your video. Music and sounds playlists are already curated in every genre, including EDM, Pop, HipHop, Rock, Country, and Rap, as well as millions of original sounds that go viral.

TikTok has all types of video ranging from Gaming, DIY, Comedy, Sports, Memes, Food, ASMR, and many more. It's an incredibly diverse media app. Your TikTok feed is also personalized based on the videos you have watched, liked and shared with other people.

To create an account:

- Launch the **TikTok** app. It's the app with the black icon and a white music note inside it.
- Select the **Create Account** option.
- You can select a sign-up method from either your phone number, email address, Instagram, Facebook, Google, or Twitter account.
- After you've chosen a sign-up method, follow the instructions that show up on your screen. The next steps may vary based on the sign-up method you chose.
- Then, select your birthday and tap the right arrow to continue to the app.

2. WhatsApp

Like other smart devices, iPhone 13 comes with WhatsApp, which is great for socializing with other people individually. It is a cheap way to stay connected internationally by sending messages, charring in group chats, and sharing multimedia files. You can also make voice and video calls.

It also works across mobile and desktop with no subscription fees and runs just as smoothly with slow connection. With more than 2 billion people from over 180 countries using this app, you can message any of your friends from all over the world.

All you need to properly enjoy WhatsApp is your phone number. You don't have to create any usernames or logins. It also has a Status feature which allows you to share what you're up to daily and keep your friends and family updated.

Your personal messages and call to your family and friends are also private to only you. No one else, not even WhatsApp themselves, can read and access these messages. This is because of the end-to-end encryption the app has which explains why it is considered one of the top messaging apps worldwide.

To create an account:

- After launching the app, **tap Agree & Continue** once you've finished reading the privacy policy and terms to start setting up your account.
- A message will then pop up asking if you want to receive notifications for WhatsApp. You can choose either **"Allow"** or **"Don't Allow."** If you change your mind later, you can go back and edit this in settings.
- Then, **type the country code and phone number** you will be using on your iPhone. When you're finished, select 'Done' found in the upper right corner of your screen.
- Once data retrieval finishes, enter your name and add a profile photo for your account. Then, select done.
- Another message will pop-up asking if WhatsApp can ask your contacts. Allowing this will sync your iPhone's contacts to the app. You can also see which of your friends and family are already using the app. After you've done this, you can then start chatting.

3. Twitter

Connecting with people worldwide and staying updated with the recent news is made possible by Twitter. From recent news, quotes from powerful people to useful business and financial tips, there is a place for everyone in this social media app.

It's the go-to social media app for staying on top of current events and a media source for everything going around the globe.

You can jump into conversations with people through Twitter Spaces. It allows you to have live audio conversations with anyone and engage in any topic.

Twitter is also a great way of finding all your interests in one space. The tweet, retweet, reply, share, and like functions you get to be more involved with people who share the same interests as you. Not to mention, it's a great way to let your opinion be heard and share any new important information and updates.

The hashtags and trending topics feature allow a more convenient way to stay in the know. When you follow your favorite influencers and content creators, you can access and stay updated with their content in just a glance.

Twitter is also a great platform for established influencers and aspiring ones to grow their community and build a more personal connection with the fans. You can easily find interesting people or build a following of people who are interested in what you do. With Twitter, you can also easily speak directly to your favorite famous people and be surprised when they answer back.

To create an account:

- After launching the app, enter your name and email address. An email with instructions will then be sent to you to verify your account.
- You can also use a phone number to create your account. If you do, a text message with a code will be sent to you to verify your number.
- You will then be redirected to your account. Here, you can tap the navigation menu and select **Profile.**
- Tap **Edit Profile** to change your display photo, your header, display name, location, birth date and bio.

- When you're done, tap **Save.**

4. Instagram

Instagram is owned by Facebook and is an excellent place to interact through visuals. Here, you can post your photos and stories and share them with the world. Instagram also allows you to follow other people and like or comment on their posts.

It's a great platform for connecting with friends as well as letting them know what you're up to. You can also stay updated and see news from different parts of the world.

You can post the highlights to your day with photo or video stories that disappear after twenty-four hours. There are also various tools you can edit them with to make them more engaging. Features such as Boomerang that allows you to loop any moment and turn it into a fun mini-video of sorts, and Superzoom where you can add special effects to make it more dramatic and entertaining as the camera automatically zooms in. It's a fun and creative way of sharing those little moments with your friends no matter how far away they are.

If you want to be more interactive with your friends and the people that follow you, you can add polls to your stories and ask silly questions. The Close Friends feature also lets you choose which friends you allow to see specific moments and highlights of your day. And, to top it all of, you can pin your favorite moments to your profile where you can constantly view them and keep them alive forever.

You can also post those really special moments to your feed. This way, anyone who visits your profile can see them too. Instagram video and reels are also fun features that lets you see more content from your favorite content creators as well as discover new ones.

You can also create your own 15-second long Insta Reel. You can bring life to it by adding special effects, face filters, emojis, background music, and stickers to make it more entertaining to watch.

The Explore page on Instagram allows you to discover new accounts, as well as small businesses and shop products that you might be interested in and would be relevant to your personal style.

To create an account:

- Launch the **Instagram** app.

- You can use either your email address or phone number to create an account. If you use a phone number, a text message with a code will be sent to you for verification. You can also select **Log in with Facebook** to create an account using your Facebook account.
- When you're done, select **Next.**
- Enter your username and password and fill out your profile info. Then select **Next** to finish setting up your account.

Top Productivity Apps

The best productivity app will let you perform tasks like editing documents and storing them safely on your device. A top productivity tool for the iPhone 13 is the Callibeo Spaces, cloudless document management and editing app that does not require an active internet connection to operate.

With this app, your team can work together on a document as long as they are on the same local network. It comes with features such as document management, document scanning, PDF viewing, text recognition, and more.

Other productivity apps include:

- **Mail**- This app has been pre-configured to work with Gmail, Hotmail, and AOL.
- **Reminders**- Get reminders of important schedules. You can use this app to write down notes and it can automatically sync to missed calls.
- **Home**- Gain control of your IoT devices in this one-in-all Internet of Things management system.
- **iMovie**- This is a video editor that comes with a timeline view.

Top Utility App

Your new device has apps designed to make your life much easier. It features utility apps that allow for more customization, turning your device into a useful digital tool. These apps include:

- **Settings**- In Settings, you get a lot of features that allow you to tweak and customize your phone.
- **Clock**- You will need this app to set the clock, an alarm, timer, or stopwatch.
- **Camera**- Keep precious memories with this app, which can also be assessed through the lock screen.

- **FaceTime**- This is a video-calling app that lets you connect with your family, friends, and business partners.
- **Find iPhone**- It is easy to find your misplaced phone with this app. The app also comes in handy when you need to remotely erase your phone in case of theft.
- **Tips**- Get tips, tricks, and shortcuts for using your iPhone to get the most out of it.
- **Measure**- This is a new app that uses AR and comes in handy for measuring dimensions like height.
- **Compass**- You can find this app useful when you need to find direction.

Top Shopping and Food Apps

Apple offers shopping and food apps for organized list makers that want to keep their shopping lists organized.

AnyList

This is a grocery list app that can help collect and organize your grocery. It has integrated recipe storage and lets you share your grocery list with family and friends. However, you will need to upgrade your account to access this feature.

Mealime

This is a free to use app that allows you to combine your recipe app with your shopping app. It lets you add ingredients to your shopping list, giving you an organized grocery list. Besides, you will get healthy, personalized recipe suggestions.

CHAPTER 6. CAMERA AND PHOTOS

Your iPhone 13 is capable of producing high-quality photos and videos no matter where you are or what time it is. It features a Pro 12MP camera system with ultra wide, wide, and telephoto cameras.

In addition, it allows you to apply effects to your videos and capture excellent shots even in low-light conditions. You can take candid pictures or use Photographic Styles to personalize your images. It also has editing tools you can use to adjust the color and light of your photos and videos, enhance, or crop.

Cameras

iPhone and iPad devices are some of the most popular still and video cameras globally. In addition to capturing stills and movies, iOS offers a powerful platform for computational photography and computer vision applications. To make the most of iPhone and iPad cameras, you should be aware of the specific capabilities of each camera device.

About The Camera Features On Your Iphone.

Learn about Photographic Styles, QuickTake, the Ultra-Wide camera, and other camera features on your iPhone.

1. **Lock In Your Look With Photographic Styles**

With Photographic Styles on iPhone 13 models, you can personalize the look of your images in the Camera app. Choose a preset, Rich Contrast, Vibrant, Warm, or Cool and if you want, fine-tune it even further by adjusting the Tone and Warmth settings. Set your style once to use your preferred style setting every time you take a photo in Photo mode.

2. **Set Up A Photographic Style**

When you open the Camera app for the first time, tap Set Up to choose your Photographic Style. Swipe through the different styles and tap Use [Style Name] on the preset that you like. Make sure to set your style before you start taking photos, you can't add your Photographic Style to a photo after you've already taken it.

3. **Change Your Photographic Style**

Want to change the Photographic Style that you set? Open the Camera app, tap the arrow, and tap Photographic Styles. Standard is the default, balanced style that's true to life and can't be customized, but you can swipe left to view other preset styles that are customizable. Tap Customize to adjust the Tone and Warmth of the style that you've chosen.

4. Capture Close-Ups With Macro Photos And Video

iPhone 13 Pro and iPhone 13 Pro Max introduce macro photography, using the new Ultra-Wide camera with advanced lens and auto-focus system for stunning close-ups with sharp focus as close as 2 centimetres. iPhone 13 Pro and iPhone 13 Pro Max can also shoot macro videos, including slow-motion and time-lapse.

Shooting macro in Photo and Video modes is automatic, just move your iPhone close to the subject, and the camera will automatically switch to the Ultra Wide camera if it's not selected while maintaining your framing. Select the Ultra-Wide camera (.5x) and move close to the subject to shoot macro slow-motion or time-lapse videos.

You might see the Camera app transition to the Ultra Wide camera as you move your iPhone close to or away from a subject. You can control automatic, macro switching by going to Settings > Camera, then turning on Macro Control. With Macro Control on, your Camera app displays a macro button when your iPhone is within a macro distance of a subject. Tap the macro button to turn off automatic, macro switching, and tap it again to turn automatic, macro switching back on.

If you turn on Macro Control, automatic, macro switching is enabled the next time you use the camera within macro distance. If you want to maintain your Macro Control setting between camera sessions, go to Settings > Camera > Preserve Settings and turn on Macro Control.

5. Grab A Video With Quicktake

You can use QuickTake to record videos without switching out of photo mode. QuickTake is available on iPhone XS, iPhone XR, and later.

Take a Photo

Your iPhone 13 comes with Photographic Styles, which you need to set up the first time you open the Camera app. Click Set Up and select your Photographic Style. Swipe the screen to choose from the available styles and click Use whatever style you've chosen.

Remember that you can only add the Photographic Style before taking pictures and not after. So you have to set it before capturing an image.

However, you can still change the Photographic Style you set. Once the Camera app is open, click on the Up Arrow and tap Photographic Styles. It is set on default, which is Standard. The default style is more true to how the image looks in real life, and you can't customize this style.

Find the other preset styles by swiping to the left. Once you find a customizable style, choose Customize and adjust its Warmth and Tone.

Using macro photo and video to capture close-ups

Both iPhone 13 Pro and iPhone 13 Pro Max are equipped with the Ultra Wide camera. It is capable of macro photography with its auto-focus system and advanced lens. You can use zoom-in as close as two centimeters from your subject and take photos with a sharp focus. You can also take macro videos that allow time-lapse and slow motion.

The macro setting for shooting Photo and Video in macro is automatic. You only have to put the phone close to what you want to take a photo of, and automatically, your phone's camera will switch to the Ultra Wide mode while keeping your framing if you did not previously choose the setting.

You can shoot a video with a time-lapse or slow-motion effect on a macro setting by choosing the Ultra Wide camera and moving closer to your subject.

You will notice a macro button, something that looks like a flower, when your phone is ready to shoot within a macro distance. You can tap it to turn it off and tap once more to turn it on.

If you want to enable automatic macro switching the next time you open your phone's Camera and place it close to a subject, go to Settings. Choose Camera, and then tap Preserve Settings before turning the Macro Control on.

Adjusting exposure and focus

The camera of your phone sets exposure and focus automatically. It also balances face exposure through face detection. To ensure that exposure and focus are set in a precise manner based on your preferences, use Exposure Compensation Control and lock the setting so you can use it on your next shots.

To adjust the level of exposure of your phone's Camera, tap the Arrow Up, then adjust the level of exposure by tapping on the Plus or Negative sign. This setting will remain until the next time you take a photo or video.

Taking photos using night mode setting

This setting allows you to take quality photos even in low-light surroundings. The feature is available in iPhone 13 Pro Max, iPhone 13 Pro, iPhone 13 mini, iPhone 13, and some other the previous models.

This setting automatically turns on when you open your phone's Camera in a low-light environment. You will notice that it's active when its icon found on top of the display becomes yellow. Taking your photo may take a while or finish quickly, depending on how dark the environment is. You can also manually adjust the setting.

To take the best photos in dark surroundings, make sure that you steadily hold your phone until it has completed capturing the image. You can make it more stable by placing the phone on a tripod or any secure and solid surface. You can also use this setting to take a selfie and capture time-lapse videos.

Taking a photo on live mode

iPhones have the 'live mode' feature where it records what happened 1.5 seconds before and after taking a photo. Live mode photos are basically just like traditional photos but you have several frame options to choose from so that you can get the best picture possible in those moments.

Here are the steps on how to enable live mode on iPhone 13:

1. Open your Camera app and go to 'photo'.
2. Tap on the circle icon found on the upper right side of the screen. There should be no diagonal line over the circle.
3. Take the photo that you want by simply tapping on the shutter button and it will be saved in live mode format.

The live mode photos are visible on the photos app. You'll know that they are in live mode because you'll see the word 'Live' on the upper left side of the screen. To preview the 1.5 seconds before and after of the photo, all you have to do is long press it.

The nice thing about live mode photos is that you can edit them in different ways such as:

- **Set a Key Photo:** You can pick another photo that you prefer better if you move the white frame in the frame view slider. Tap on the photo that you want and press done.
- **Trim Live Photo:** Drag the slider to the end of the live photo frame and leave out the part that you want to trim.
- **Make a Still Photo**: You could also make a single still photo with this feature by tapping on the live icon and turning it off. The key photo will then become a still photo.
- **Mute Live Photo:** Live photos are like video in a sense that the audio is also recorded when taking the picture. If you want to remove this, you just need to tap on the music volume icon.
- **Add Effects**: You can also add effect to the live photos by tapping on the live icon found on the top-left side of the screen. The effets option you can choose from include:
    1. Long Exposure
    2. Bounce
    3. Loop

Record a Video

To record a video, open the Camera app and change the mode selections based on your preference.

Start taking a video

First, choose Video and tap the button that indicates Record. You can also start video recording by pressing the volume button. You can perform the following while the recording is ongoing:

- Hold 1x to get a precise zoom, and move the slider to the left
- Pinch your screen to zoom out or zoom in
- Take a still photo by pressing the white Shutter button
- To stop recording, you can press either the volume button or the Record button.

The default setting in recording a video is 30 frames per second. If you want to record at faster frame rates, you can adjust the resolution and frame rates by choosing Settings, pressing Camera, and tapping on Record Video. Note that the higher the resolution of your video, the more storage it will consume.

Record in Cinematic mode

This mode makes the background and foreground blurry while making the subject sharp to create a depth-of-field effect. When you have this on, your phone will continuously detect the main subject and keep the focus on it until the end of the shooting. The phone will transition to another subject automatically if it identifies a new point of focus. You can also choose the subject of the video by manually adjusting the setting.

To apply the setting, choose the Cinematic mode. Before recording, tap 1x to zoom closer. You can tap the button with an F logo to adjust the effect of the depth of field and hold the slider right or left before you begin recording.

You can again choose to press the volume button or the Record button to begin recording. You can check the subject in focus because a yellow frame indicates it. On the other hand, the gray frame appears when another person is on the screen but is not part of the focus. You can change the focus by tapping on the gray box and lock the selection by tapping on the gray box one more time.

If you are not shooting a person, you can tap anywhere on your phone screen to choose what you want the camera to focus on. You can touch the screen and hold it when shooting at a single distance, and you want to lock the focus. Tap the volume button or Record button to stop recording.

You can change or remove the effect after recording the video in this mode.

Record video in a slow-motion setting

This setting will record video like normal but plays with a slow-motion effect. You can also edit when the slow-motion effect starts and resume within the video.

First, tap on the Slo-mode. In iPhone 13, you can do this using the front camera by tapping on the logo for the setting. You can then press the volume button or the Record button to start recording. Your phone allows taking a snap while the recording is ongoing by tapping on the Shutter button. Stop recording the video by pressing the volume button or the Record button.

If you want to choose a specific point in the video to be played in a slow-mo while the rest plays at normal speed, choose video, and tap Edit. Drag the vertical bars under the frame viewer to start selecting the part of the video that the slow-mo will be applied to.

Recording video on a time-lapse setting

This is typically used in capturing the sun shining or setting or the flow of traffic. To get this done, choose Time-lapse, and make sure that your phone is placed on a solid surface where there will be minimal or no interference during the recording. Start recording the video, come back when your intended time duration is finished, and then stop recording.

Record on ProRes

As you already know, iPhone 13 has the ability to shoot high-definition video because of its 4k resolution capabilities. However, the iPhone 13 Pro Max takes it to another level with its ProRes video capabilities.

ProRes is Apple's video codec which allows its camera to capture more information when capturing videos. This allows for more control in post production when using software like DaVinci Resolve, Final Cut, and Adobe Premier.

In the past, ProRes features were only used by video production professionals. But, Apple has made it readily available with the iPhone 13 line.

Here are the steps on how to use it:

1. Make sure that your iPhone is on the latest iOS version.
2. Open your settings app.
3. Scroll down to the camera settings. You can also type in 'Camera' on the search bar.
4. Tap on 'Formats'. Under its subfolder, you'll find the toggle for the ProRes video capture. Slide it to green so that it's switched on.
5. Go to your camera and tap on 'video'. You'll now see a ProRes icon besides the HD option on the upper right side of the screen.
6. Tap the ProRes icon to activate and the white lines on the screen will be gone.
7. Press the record button and your video will be recorded on ProRes formal.

This is a great video format if you want to make the best out of iPhone camera quality and you plan on doing plenty of post-production work. However, take note that this format can consume a lot of memory space.

In fact, a one minute long ProRes 4k video is around 6GB compared to a one minute long 4K video which is only 200 MB. Even if your iPhone has plenty of available memory space, it's best not to fill your storage quickly so that it won't affect the overall performance of the phone.

Also, take note on the entry-level iPhone 13, the ProRes recording capacity is limited to 1080p instead of 4k.

Print Photos

To print photos from your phone, first, you have to open the Photos app. Choose the image that you want to print, and click next. Press the Share button, which you can also use to choose multiple pictures that you want to print.

You will find the Printer icon between the Assign to Contact and Slideshow. Tap on it. Choose Printer Option, then Select Printer. Your phone will connect with your Printer. If you have done this before and use the same printer, it will automatically connect to it. Choose the Print button, and your photos will begin printing.

Edit Photos

Aside from getting more cameras and improving their performance, this unit also features a more powerful photo and video editing capability. It has an updated layout, designed to make it easier for you to edit with your fingers.

Here's the steps on how to edit photos and video using the iPhone 13's photos app:

1. Launch the Photos app and tap on the photo or video that you want to edit.
2. Tap on 'Edit' which is found on the upper right corner of the screen. You'll see three main editing options on the bottom of the screen; adjust, filters, and crop.

   Under the 'Adjust' option, you'll find a wide selection of editing tools that will enable you to get the image that you want. Here are of the editing tools that you'll find under this option:

- **Auto:** This is the quickest way to edit and improve photos and videos on an iPhone13. Simply tap on it and it will automatically sharpen, boost color, and do other intelligent changes on your photos and videos.
- **Exposure:** You can use this to control the light in a photo. You can make an image look brighter or darker by moving the slider on different ends.
- **Brilliance:** This tool changes the contrast to highlight the fine details in the photo and make it look brighter.
- **Shadow:** Shadows functions in the same way with brilliance except that it darkens the shaded areas of the image to add more depth.
- **Contrast:** The contrast tool enables you to add a wider color range between the darker and lighter areas of the photo. Most of the time, images with higher contrast appear to be more vibrant. However, too much contrast can make a photo look bad.
- **Saturation:** You can use this tool to decrease or increase the color intensity of the photos.
- **Vibrance:** You can boost your photo's color intensity with this tool since it allows you to push the image color to the warmer end of the spectrum.
- **Tint:** This tool helps in shifting the color balance of a photo from a purple hue to green or somewhere in between.
- **Sharpness:** This tool allows you to bring focus to your photo by reducing blurriness but over-using it can cause graininess.
- **Definition:** This tool is helpful in defining the focus of your photo by bringing the parts unnoticed into view.
- **Noise Reduction:** You can use this to reduce the camera noise, removing the small dots that are visible on your photos when you zoom in.
- **Vignette**: This allows you to darken or lighten the frame to add a more vintage feel to the image.

3. Tap on 'Done' to save the photo or video you edited.

Share Photos and Videos

You can share photos and videos from your Messages, Mail, or Photos app. This is done through the following:

- To share videos or photos from a month or day, select Library, choose Months or Days, tap the logo with three dots, and click on Share Photos. You can choose to share all the pictures in that month or day.
- To share multiple videos or pictures, open the Library, choose All Photos, and press Select. Choose the images you want to share, then tap the Share button and select your preferred option for sharing.
- To share one video or photo, open it and tap the Share button before choosing your preferred option for sharing.

You can also share iCloud Photos. Once done, the link will be available for 30 days, which anyone can share and view via Mail or Messenger.

Scan QR Code

You don't have to download any app to scan and read a QR code when using an iPhone 13. The app is already included in its camera's software.

To do this, open the Camera app. Using the rear camera of your phone, point it to the QR code and make sure that the code is bound in the yellow square and looks clear. Your phone will show a notification indicating that it has read the code. You will usually be redirected to the web address the code contains when you click on it. Your phone will automatically open your default web browser.

Music and Video

Your iPhone 13 will keep you entertained even when you're not exchanging messages or calling anyone. It has apps and sites where you can listen to music, watch videos, purchase media, and get updated about news and current affairs.

Listen to Music

Open the Music app to start playing music on your iPhone 13. It will allow you to repeat songs, shuffle, skip, pause, and play. You can also view album art by viewing the Now Playing feature.

Once you are on the Now Playing view, you will have additional controls, which include the following:

- Drag the playhead to get to any point you like in a song
- Choose the artist, playlist, or album by navigating their categories
- Drag the slider intended for the volume to adjust or use the volume buttons found on the side of the phone

To check the time-synced lyrics available in many Apple Music songs, tap the button with a logo that looks like a double quote inside a dialogue box. You can access the button as long as you are in the Now Playing view. After tapping, you will see the lyrics scrolling as the music plays. Tap the logo again if you want to hide the lyrics. You can drag the slider to any point of the song if you're going to get its lyrics. You can also tap the logo with three dots to check the Full Lyrics. Note that viewing lyrics is available to Apple Music subscribers.

You can also stream music to devices that enable AirPlay or Bluetooth. Go to Now Playing, choose the logo with what appears like a Bluetooth signal, and choose the device where you will stream.

This feature will allow you to play the same music on different devices simultaneously or share the music you are playing to other iPhones or AirPods.

Watch TV and Music

You can use the FaceTime app to stream music, movies, and TV shows simultaneously with family or friends via SharePlay. You can all watch it together during your FaceTime call. Everybody connected in the call can access shared controls and synced playback. The volume of the media being played can also be dynamically adjusted using the smart volume feature. This will let you all chat as you listen or watch.

To start watching videos together on FaceTime, first, start the call. Open the video streaming app on your Home screen. Note that the app has to support SharePlay. Choose the movie you will watch, choose Play, and then Play for Everyone.

Everybody in the FaceTime call can also listen to music together. To do this, first, start the call. Open a music streaming app on your Home screen, and make sure that it supports SharePlay. Choose the music and press Play. This will share the music with everyone in the call who has access to the content.

Purchase Music, TV, and Movies

To purchase media content on your iPhone 13, go to the iTunes Store app. You can then choose any of the following:

- Search - type what you want to watch or listen to and tap Search
- Charts - check out the popular media on iTunes
- Browse by category - you can browse by categories, including TV Shows, Movies, or Music. You can also tap Genres to refine your search
- More - you can go through the recommended list based on your past purchases from iTunes

Once you have chosen a media file, tap it to see more details about the item. You can choose to watch trailers of TV shows or movies or preview songs. Once you have selected, you can perform the following actions:

- Add to wishlist - you can add the items to your wishlist and go back to them later
- Tap gift - you can give the chosen item to other iPhone users as a gift
- Share - you can send your friends the link to the media file

Select the price or tap Get if the item is free when you are ready to buy and download the content. If you have purchased it before, you no longer have to pay for it, but you can download the item again.

There are instances when you will be required to show authentication proof, such as your passcode, Touch ID, or Apple ID with Face ID, to complete the transaction. You can check the progress of your download by tapping the More button and Downloads.

Your phone also allows you to send or redeem an iTunes or App Store gift card. You only need to tap Music, go to the bottom of the screen, and press Send Gift or Redeem.

Get the Latest News

Your iPhone is equipped with the News app, where you can read and share the latest happenings from around the world. You only have to tap your chosen story to read it. While reading, you can perform any of the following:

- Tap Arrow Right at the lower right part of your screen to read the next story
- Tap Arrow Left at the lower right of your screen to read the previous story

- Tap the Like or Dislike button to tell News which stories you prefer reading
- Swipe from the left corner of the screen to return to the list of news you can view
- Tap Share Story to share what you are reading via Mail or Messages
- You can make the texts larger or smaller
- Tap Report a Concert if you see anything offensive, inappropriate, or mislabeled content

Web and Communications

You can keep yourself updated and connected while using your iPhone 13. It has web browsers, enables chatting and calling through Messages and FaceTime, and allows you to connect with other devices. You can also use the gadget to organize your files and check or manage your schedule.

The Browsers

You can browse the web using your iPhone using its default browser, Safari, or by installing Google Chrome. To start browsing, tap your chosen browser from your phone's Home screen. You can enter keywords you'd like to read about at the bottom part of the browser or input the URL of the website you want to visit.

In Safari, you can turn the Quick Website Search on at the app's Settings. It will allow you to quickly search for the sites or keywords you have previously visited. If you want to bookmark a site for easy navigation the next time you use the browser, choose the Menu button and Add Bookmark. You can select a name for the site you have bookmarked before clicking Save. You can access links on your Bookmarks by tapping the Bookmarks icon on the search field at the page's bottom and choosing the site you have bookmarked and want to visit.

Here are the other options you can perform using a web browser:

- Choose the Tabs icon to access Browser Tabs
- Tap the Plus or Add icon to open a new tab
- Tap the X icon or swipe the screen to the left to close a tab
- Hold the link and choose Open in New Tab to perform the action

You can also refresh a website in times when it lags or gets stuck by tapping the Menu and choosing Reload. Additionally, you can add sites to your Favorites list by tapping Menu and Add to Favorites. You can also rename the site before saving.

Messages

The Messages app on iPhone allows you to connect to other people by sending unlimited documents, videos, photos, and texts to any watchOS, macOS, iPadOS, or iOS device through iMessage.

Here's how to set up iMessage on your iPhone 13:

1. Connect your iPhone 13 to an active web connection.
2. Launch the settings app.
3. Scroll down and look for 'Messages'. Tap on the icon to open the rest of the options.
4. Scroll down and look for 'iMessage'. Flip the switch to turn on the iMessage feature.

It might take several minutes for the iMessage feature to activate but those are the quick and easy steps on how to set it up. But, if you want to check if it's working already, you can do this:

1. Open the message app on your iPhone13.
2. On the top right corner of your screen, tap on the compose icon.
3. Look for a person who has an active iMessage feature. You'll know if their iMessage is active if their name turns blue.
4. Send a message to that person. You'll know whether your iMessage is activated if the bubble is blue instead of green.

Aside from the iMessage feature, you can also make the best out of your iPhone 13's messaging app by adding an Emoji keyboard. Here's how to do that:

1. Open the settings app.
2. Go to 'General' then look for 'Keyboards' and tap on 'Add New Keyboard'.
3. There you'll find the 'Emoji' option. Tap on it to enable the keyboard.

After setting up the emoji keyboard, you can now use these emoticons on your messages. Here are the steps on how to use it:

1. Open your message app.
2. Tap on compose.
3. On the keyboard, you'll find a smiley face on the left side of the space bar. Tap on it to see the list of emoji.
4. Select the emoji that you want to use for your conversation.

5. You can also change the skin tone of the emoji. To do this, simply long press your chosen emoji. A selection of different skin tones will appear. Tap on your desired color.

Sure, using emoji is cool but Apple has stepped up its game with Animoji and Memoji. These features superimpose over your face to make messaging and video chat more fun. Here's how to use the Animoji Camera effect on your message app:

1. Open your message app. Tap on the camera icon found inside the conversation just above the keyboard.
2. Tap on the star-shaped icon.
3. Tap the Monkey button to get the live, animated effects. If you want to use Memoji, tap on the icon featuring 3 heads.
4. The camera screen will appear. Take your pose or record your video.
5. After that, send it to your recipient of choice.

Take note that this feature is only possible for two people who have active iMessage features on their iPhones. If you are planning on sending it to someone who's on Android, the message will not push through.

Aside from adding emojis, animoji, and memoji on your iPhone 13's message app, you can also do these other things:

Pin Conversation

This feature makes it super easy for you to access your conversation with someone you frequently send messages with. You can set upto 6 pinned conversations so you don't have to spend time scrolling your inbox and look for a particular person.

You can pin a conversation using these three methods:

Method #1:

1. Swipe right on a conversation
2. Tap the pin icon

Method #2:

1. Long press a conversation
2. Tap on the pin icon found on the pop-up menu.

Method #3:

1. Long press on a conversation.
2. Drag it into the top of the message screen until it turns into a circle. Drop the icon.

Reply in Line

You can now reply in line on the iPhone message app like how you would on other social messaging apps. This helps prevent conversations from getting muddled with new conversation, mitigating the risk of miscommunication.

Here's how to reply in line using your iPhone 13:

1. Long press on the particular message that you want to reply to.
2. Type in your response.
3. Send

The receiver will see that your message is a direct reply to that particular line.

Set up Group Photo and Name

The new iPhone message app allows you to change the group name and photo for more personalization. Here's how to do that:

1. Open the info of the group
2. Tap in 'Change Name and Photo'.
3. Type the name that you want for the group then tap the camera to add an existing photo from your camera roll or take a new one.
4. Tap on 'done' to save the changes that you made.

In addition, just like in social media messaging apps, you can also now mention someone by name through the iPhone message app. All you have to do is to simply type '@' + the person's name and it will turn into a mention.

Conversely, you can also get a notification if you were the one mentioned in the conversation. Here are the steps on how to enable that feature:

1. Open the 'Settings' app.
2. Go to 'Messages'.
3. Look for 'Mentions'. Turn on the option 'Notify Me,'

By doing this, you'll be notified everytime a group conversation mentions your name. It also works for muted conversations.

Filter Unknown Senders

Getting messages from unknown senders is annoying to say the least. Luckily, you can prevent clutter on your iPhone 13 messages by filtering them out. Here's how to do that:

1. Open the 'Settings' app.
2. Go to 'Messages'.
3. Look for the message filtering section and switch on the option for 'Filter Unknown Senders'.

FaceTime

This app allows you to make audio or video calls on your iPhone 13. First, make sure that the app is activated. Choose the Call icon or Phone app, add the desired numbers in your Contacts list, and choose the Phone or Video icon on FaceTime to start the call.

You can also make a FaceTime call by choosing the FaceTime app on your phone's Home screen, selecting the contacts you want to call, tapping Call or New Face Time, and following the succeeding prompts. You can choose to Create Link to send the FaceTime link to Windows or Android users. To answer a FaceTime call, you only need to select the designated icon for Accept or Decline.

Mail Account

You can manually or automatically set up an email account on your phone's Mail app. For the automatic option, choose Settings, go to Mail, and tap Accounts. Choose your email provider when prompted to Add Account. Input your email and password, select Next, and verify your email if prompted. Click Save, and you're done.

To set up your email manually, go to Settings, Mail, and choose Accounts. Tap Other once you are already on the Add Account option. Choose All Mail Accounts. You will then provide the email address, your name, description of the account, and password. Click Next and Done.

Connect Your Phone with Other Devices

The nice thing about the iPhone 13 is that it fits into the Apple ecosystem, allowing you to have continuity with your other devices. You can easily pair it with other Apple devices. Here are the steps on how to do that:

iPhone 13 & Mac

1. Open the settings app on your iPhone 13.
2. Go to 'Bluetooth' and turn it on.
3. Open your Mac and click on the Bluetooth icon found in the menu bar.
4. Go to 'System Preference' then select 'Bluetooth'. There, you'll find your iPhone 13 name.
5. Click on your iPhone 13 name.
6. You'll get a connection request on your iPhone. Tap on connect and the two devices will be paired with each other.

You might need to provide a confirmation code on your first attempt to connect an iPhone 13 with a Mac. Simply follow the instructions that will appear on your screen and input the code.

iPhone 13 & iPad

1. Open 'Control Center' on your iPhone 13 or iPad.
2. Long press on any of the icons on the box until the toggles appear.
3. Long press on the Bluetooth image. You'll be given a list of available Bluetooth-enabled devices.
4. Select your iPhone or iPad. Wait for several seconds and the devices will be paired with each other.

iPhone 13 & AirPods

1. Go to your iPhones Home Screen. Make sure that your Airpods' charging case is open and close to your mobile device. In a few seconds, an Airpod set up animation will appear on the iPhone.
2. Tap 'Connect'. If you have an AirPods 3rd generation or an AirPods Pro, read the three screens that will appear.
3. Tap 'Done'.

Pairing your AirPods with your iPhone will automatically support other devices that are linked into the iCloud account that's registered in the iPhone. This will make pairing your Airpods with other Apple devices much easier.

In addition, you can use 'Hey Siri' immediately after pairing your AirPods with the iPhone if it's already set up beforehand. But if it isn't, then you'll see a guide on how to do that. Simply follow it if you want to enable the feature.

iPhone 13 & Apple Watch

1. Wear your Apple watch comfortably on your wrist. Make sure that the band size is comfortable and secured.
2. Turn on the Apple watch by pressing and holding the side button until the Apple logo appears.
3. Place the Apple Watch near your iPhone 13 then tap 'Pair New Watch'.
4. Tap on 'Set Up for Myself'.
5. Follow the instructions that will appear on your iPhone 13 and Apple watch to finish the set up.

Once the two devices are paired, you can tap on ' Get to Know Your Watch' found on the iPhone. This will inform you with the features of the Apple Watch and how you can optimize it with your phone.

Managing Files

You can securely access and manage your documents on your iPhone 13 with the use of iCloud Drive. You can access this by signing into iCloud. The storage limit of the iCloud Drive is 50 GB but if you need more, you can upgrade it into iCloud +.

Here's how to access your iCloud Drive files on your iPhone:

1. Sign into your iCloud account that has the same Apple ID with your other devices.
2. Go to the files app. There you'll see the files and documents that you have saved on iCloud Drive.

You can also access iCloud Drive files even when you're offline. Follow these steps:

1. Open the files app.
2. Select the file that you want to download into your iPhone
3. Tap on the file or touch and hold until the download button appears.

The changes you made on the document or file while working offline will be automatically uploaded to iCloud Drive once you are reconnected to the internet.

Suppose you accidentally deleted a file and you want to recover it. Here's how to do that on your iPhone 13:

1. Sign in to your iCloud Drive Account
2. Launch iCloud Drive
3. Tap on 'Recently Deleted Items' found on the bottom right corner of your screen. Another option is for you to go to settings then select 'Restore files'.
4. Look for the file that you deleted. It should be there if the deletion is within the 30-day hold period.

Manage Your Calendar

You can choose the Calendar app on your iPhone 13 to edit and create meetings, appointments, and events. To make it easier, instead of manually organizing and editing the Calendar, you can ask Siri to set up a schedule instead. You can say something like, "Set up my meeting with Andrew at 8," or "Where is my 4 PM appointment?"

You can also follow the prompts on your screen to add an event, add an attachment, edit or delete an event, add an alert, or find events in other apps using the Calendar app.

CHAPTER 7: UTILITIES AND MAPS

Use Maps to Navigate the world

The maps app in iPhone 13 can help users get driving directions. Like other Apple services, Apple map is an exclusive Apple app. The map offers an easy to use interface that keeps everything in one place.

Viewing Maps on iPhone 13

Opening the Maps app , allows you to find your location. You can also zoom in and out to spot any details that you might want to see.
The iPhone 13, in particular, is one of the supported iPhone models that can provide enhanced detail for roads, trees, buildings, and more in selected cities on Maps.

Using precise location on Maps

Connect the iPhone to the internet and turn on Precise Location in order to find your location and be provided with accurate directions.

If a message pops up saying Location Services is off then Tap the message, then tap Turn On in settings and turn on Location Services.

If a message pops up saying Precise Location is off then Tap the message, then tap Turn On in settings, tap Location and turn on Precise Location.

Note that cellular data rates may apply.

Tip: Go to Settings > Privacy > Location Services > System Services then turn on Significant Locations to get more useful location-related information in Maps.

Showing your current location.

To see your current location, tap .

Your current location is marked in the middle of the map. The top of your map shows north. If you want to see your heading instead of north, tap . To resume showing north, tap or .

Choosing the right map

The top right of the map shows a button that serves as indicator whether the current map is for driving , riding transit , exploring , or viewing from a satellite .

You can switch between the different maps by tapping the button at the top right and choosing the desired map type, then press ×.

Viewing a 3D Map

When on the 2D Map interface:

Drag two of your fingers up.

Tap the 3D button near the top right on the Satellite Map.

When on the 3D Map interface:

Drag two fingers to adjust the angle.

Zoom in to see buildings and other details.

Tap the 2D button near the top right if you wish to return to the 2D map.

Searching for a place

You can ask Siri, when searching for a place, by saying something like "Show me [name of location]."

Otherwise, you can tap the search field, which can be found at the top of the search card, and type in your desired location.

You can find the location by typing any of the following:

Area

Landmark

Intersection

Business

Zip Code

If you get more than one results, scroll the list to see more. Then, tap a search result to learn more about it or to get directions to it.

Reminders

The Reminders app allows a convenient to make to-do lists for different tasks and projects. You can also create subtasks, add attachments, set flags, and several other features. Reminders also allow you to set alerts based on location and time.

Tip: Use upgraded reminders to have all the Reminders features available for you. Using other accounts might also cause some features to be unavailable.

Creating a list

To create a new list:

Tap Add List. For those with multiple accounts, choose which preferred account to use after adding list.

Enter your name, then choose a preferred color and symbol for your new list.

Adding a new reminder

You can add a new reminder by telling Siri something along the lines of "Add [name of item] to my [name of list]."

Otherwise, you can do the following:

Tap the New Reminder button, then enter your text.

You can use the buttons above the keyboard to:

Schedule date and time: To do this, tap , then choose the preferred date for the reminder.

- Add a location: You can choose where you want to be reminded by tapping the icon.

Tip: Precise Location must be turned on to receive location-based reminders. Turn on Location Services in the Settings, tap Reminders, choose While Using the App, then turn on Precise Location.

- Assign the reminder: This feature is available in shared lists. Tap the icon, then choose which person to assign the reminder to. You can also choose yourself.

- Tap ⓘ if you want to add more details to your reminder, such as notes, web link, and a priority level.

If you have nothing more to add, tap Done.

Creating a subtask

In most cases, reminders have subtasks that include things that you don't want to forget. There are different ways of creating a subtask including:

To create a subtask, swipe right on a specific reminder. Then tap Indent. You can also do this by dragging a reminder onto another reminder.

You can also tap and hold a reminder before dragging it to another reminder.

The other method is to tap a reminder and click on the Edit Details button. Then click on Subtasks followed by Subtasks and type your subtask. You can do thing as many times as you would like to create subtasks.

Completing the main task also completes the subtask. Similarly, deleting or moving the main task also deletes or moves the subtask.

Edit a Reminder

You can modify and customize a reminder based on your preference. Click on the Edit Details button. This will give you the option to customize and add more details to your reminder.

Besides customizing the reminder, you can also tweak its notification settings, including the list where it belongs to.

Edit Multiple Reminders

It is possible to edit multiple reminders all at the same time. Here is how to do this:

Open the list and tap the More button.

Click Select Reminders and choose the reminders you want to edit. The other option is to drag two fingers on the reminders you want to edit.

Navigate to the bottom of the screen and click on the options provided to perform an action on the selected reminders. Here, you can complete, move, flag, assign, delete, or add a date and time.

Completing a Reminder

Once you finish a reminder, you can mark it as complete and hide it by tapping the empty circle right next to it.

If you want to unhide the completed reminders, tap ⊙, then tap the Show Completed option. Tap the Clear option if you want to delete the completed reminders.

Adding a Reminder with Siri

iPhone 13 gives you the opportunity to schedule a reminder with Siri. For instance, you can as Siri to remind you to check your mail at a specific time. To make this more efficient, consider adding your addresses to the card in your Contact, allowing Siri to set the reminders based on your location.

Reminder When Messaging

Your iPhone device also lets you set a reminder that will pop up when chatting with someone.

Tap a reminder and click the Edit Details button.

You will see the option to turn on/off When Messaging. Turn it on.

Click on Choose Person and select a name from your contacts to add

To do this:

Launch the Contacts app.

Click on My Card and the tap Edit.

Enter your home or work address and tap Done.

Deleting a Reminder

If you don't want to schedule your tasks with a reminder anymore, you can do so by deleting the reminder. To do this, go to the reminder and swipe left. Then click Delete.

However, if you change your mind about deleting your reminder, you can easily recover it. To undo the deleting process, shake or tap with three fingers.

Adding a Reminder from Another App

If you want to remind yourself to revisit another app, you need to add a link in your reminder. Open the website or app that you want to revisit and locate the Share button. Find the Reminders icon and click on it.

Calendar

iPhone 13 has a Calendar app that you can create and edit appointments, meetings, and events. This calendar helps users stay organized and makes it easy to track events and appointments.

Creating and Editing events in Calendar

You can create and edit events in the Calendar app by asking Siri something like:

"Set up a meeting with [name] at [date]"

"Where is my [time] meeting?"

"Do I have an event to attend to at [date and time]?"

You can also do the following:

Adding an Event

- When in day view, tap + that can be found at the top of the screen.

Enter the name or title of the new event.

If it is at a physical location, then tap Location. For a remote event, tap FaceTime to enter a video link.

Enter all the necessary details for the event, like start and end time, travel time, invitees, etc. You can swipe up to enter all the meeting information.

When you're done, tap Add.

Adding an Alert

You can choose to set an alert as a reminder for a specific event by doing the following:

Tap the specific event you want to be reminded of, then tap Edit found at the top right.

Tap Alert in the event details.

Choose the specific time you want to be reminded of the event.

Tip: Adding the address of the event's location enables Calendar to use Apple Maps to look up traffic conditions and transit options, then alert you of the appropriate time to leave.

Editing an Event

You can edit any event details like the date and time by doing the following:

In day view, touch and hold the event, then drag it to the new preferred time to change the time of the event.

Tap the event you want to edit, then tap Edit found near the top right. In the event details, tap any setting you wish to change, or add a new field to type new information.

Check Multiple Calendars at Once

You don't to keep track of all your meetings, appointments and event in one place. You can also create additional calendars for different events. From your list of Calendars, you can choose which calendars you want to view.

You can also include birthdays and national holidays with your events. Besides, there is also the option for Siri suggestions.

Set a Default Calendar

When you set a default calendar, all events added using Siri or other apps will be included to your default calendar. To set the default calendar:

Launch the Settings app.

Click on Calendar.

Next, tap Default Calendar and choose your default calendar.

Subscribe to a Calendar

To subscribe to a calendar:

Open the Settings app and click on Calendar.

Tap Accounts and click Add Account then click on Other.

Click on Add Subscribed Calendar.

Input the URL of the file calendar you want to subscribe to.

If you change your mind, you can unsubscribe you a calendar.

Navigate to the bottom of the screen and click on Calendars.

Click the (i) icon next to your target calendar.

Click Unsubscribe to remove the calendar that you subscribed to. You can also click on Unsubscribe and Report Junk to report the calendar as junk subscription.

Move the Calendar's Color

Navigate to Calendars at the bottom of the screen.

Tap the (i) icon next to the calendar and select Color.

Tap Done.

Clock

The date and time on iPhone devices are set by default depending on your location. Nonetheless, you have the option to adjust the date and time on your device.

Features of Clock App in iPhone 13

iOS 13 comes with many upgrades and this is no exception for the clock app. Some of the new features to expect include:

Different Bedtime Schedules

Bedtime schedule was first introduced in Apple's iOS 10 version and is designed to help users record when they slept and when they woke up. In iOS, you can set different bedtime schedules for different days to serve as your morning and bedtime alarm.

Time Zone Notification

You will be notified when you are in a new time zone. When you get this notification, you can make the necessary adjustments so you can have updated information.

Pre-Alarm Notification

The pre-alarm notification is silent and it goes off several minutes before the scheduled time. This way, you can dismiss the upcoming alarm before it wakes up.

Syncing Apple Watch and iPhone Clock

Both the Apple Watch and iPhone clock can be synched together so that they can share timers and alarm that are lacking in the other. The synching can be done directly or via iCloud.

Changing the Date and Time

Open the Settings app.

Tap on General and then click Date & Time.

To change the default time and date, switch off Set Automatically and turn on 24-Hour Time.

Adjust the date and time depending on your location.

Accessing the World Clock

You can use the Clock app to view the local time in different time zones all over the world.

Ask Siri something like "What time is it in [name of a place outside the country]?"

You can tap the World Clock, then manage the list of cities by tapping the Edit button, then adding a city by tapping +, or deleting a city by tapping ⊖.

9.4.3. Adding Digital Clock on Widget

There is no functionality to add a digital clock on widget and you need to use a third-party app.

After downloading a third-party app, launch it and create a digital clock widget.

Long press an app icon and tap on an empty area on the screen to launch the Home Screen edit mode.

Navigate to the + icon on the top-right corner of the screen and click on it.

Locate the app name that you have used to build the digital clock widget.

Add the right glance and long-press the widget.

Tap Edit Widget and select the digital clock widget that you had built earlier.

Setting an Alarm

You can set up a regular alarm for any time of the day, unrelated to your sleep schedule, by tapping +.

You can set a wake up alarm by tapping Alarm, then tap Change. Then adjust your wake up time by dragging and your sleep time by dragging .

Notes

Creating a new note

To create a new note in the Notes app , you can ask Siri to do so or by doing any of the following:

Tap , then enter the text for your new note. Keep in mind that the first line of your note becomes the note's title.

You can change the formatting by tapping Aa.

Tap Done to save the new note.

Tip: You can choose a default style for the first lines in any new notes you create by going to Settings > Notes > New Notes Start With.

Adding a checklist

To add a checklist, tap . Then, you can do any of the following:

Adding items in the list by entering the text and tapping return to enter the next item.

Swiping right or left on a specific item to increase or decrease the indentation.

Tapping the empty circle next to the completed item to add a checkmark.

Touching and holding the empty circle to reorder the item in the list.

Creating Note from Lock Screen

The Notes app allows you to create your note from the lock screen using an Apple pencil.

Go to the Settings app.

Click on Notes.

Click on Access Notes from Lock Screen and select an option.

Once done, use your Apple Pencil to tap the lock screen and create your note. This will then be automatically saved in Notes.

Deleting Notes

Follow the steps below to delete Notes from your Apple device:

Open the note and click the three dots in a circle and tap Delete.

Alternatively, go to the note that you intend to delete and swipe left over it.

Click on the Trash button to delete it.

Recovering Deleted Notes

If you have deleted notes from your iPhone 13, you can retrieve them in simple steps.

Open Notes and click Recently Deleted.

Click on the three dots in a circle and click on Select Notes.

Tap Move To and create or choose a folder to save the notes.

Setting Up Notes with iCloud

If you want to update your note on various devices, you will need to set up your notes with iCloud. Here is how:

Open the Settings app and then go to your name.

Tap iCloud and turn on notes. Here, you will see your notes on devices that you have signed in using the same Apple ID.

Disable Notes from iCloud

iCloud can be used to back up all your data including Notes app. However, you may need to disable Notes from iCloud, especially if the Notes app is crashing on your phone. Here is how to:

Open the Settings app and click on Initials.

Tap iCloud Option followed by Notes app.

Turn off the toggle to disable Notes from iCloud.

When asked to confirm whether you want to delete the Notes app, confirm Yes.

How to Update to the Latest Version of Notes App

If you are using an outdated version of Notes app, you may encounter issues like the app crashing. To update your Notes app to the latest version:

Open the App store and search for any update on Notes app.

If an update is available, click on it.

How to Format a Note

You can format your notes by adding heading, title, bullet list, or a table.

Open the note and click on the table or formatting button (Aa).

Format the note by adding a title, table, bulleted list, or heading.

How to Pin a Note

When you pin your notes, it will be easy for you to retrieve them.

Swipe right over the note you intend to pin before releasing.

Another method to pin your note is by tapping the More button (three dots in a circle) and click on the Pin button.

If you want to unpin a pinned note, go to the note and swipe right over it.

How to Add an Attachment

You can also add an attachment to your note from another pin. Here is how:

Open the app you intend to share from and click on the Share button.

Click on the Notes app and choose the note you want to include an attachment to.

Once done, click on Save.

If you are looking for a note or attachment, click on the Search field and type what you are looking for. You can also search for images include your notes or attachment by typing the name of that image.

When searching for an attachment, click on the More button with three dots in a circle and click on View Attachments. Open the note that contains the attachment. Then, touch and press the attachment's thumbnail and then click Show in Note.

How to add a Photo or Video

You can add a new or existing photo or video to your app.

Open a note and click on the Camera button.

You will be provided with the option to take a photo or video, choose an existing one, or scan documents. Choose your option to add a photo or video to your notes.

CHAPTER 8: HEALTH AND FITNESS

Using an Apple Watch

Key Parts of the Apple Watch

Although an Apple Watch may appear small, it has many features and functions that you must know for a seamless user experience. Some of the key parts to familiarize yourself with include:

Display- This comprises of the face and screen of the Apple Watch where you can tap, swipe, and press to perform different functions.

Digital Crown- The second part is the rotating button that is located on the side of the watch. You can press this button to see the Home screen or the Watch face. Turning it allows you to zoom or scroll while pressing and holding it allows you to use Siri. If you want to open the last app that you used, double press the Digital Crown.

Side Button- Right after the Digital Crown is the side button. When you press or hold the button, it responds by turning your watch on/off. Performing this action can also allow you to make an emergency phone call. Press the button to see your recently used apps and double press to open Apple Pay.

Band Release Buttons- The Apple Watch bands can be removed thanks to the band release buttons at the back of the watch. Use the buttons to remove your current bands and install new ones.

Optical Heart Sensor- At the back of your Apple Watch are four dots that you can use to measure your heart rate.

Pairing Apple Watch with your iPhone

To pair your Apple Watch with your iPhone, tap the icon, then follow the instructions displayed on your screen.

Unlocking iPhone with Apple Watch

You can use Series 3 and later Apple Watch to unlock your iPhone with the Face ID feature. To allow your Apple Watch to do this:

- Go to Settings > Face ID & Passcode.

Scroll further down the list, then turn on Apple Watch.

To unlock your iPhone when you're wearing your Apple Watch and a face mask, raise your iPhone or tap its screen to wake it, then glance at your iPhone.

Important: Your Apple Watch must have a passcode. You must also be wearing it on your wrist and have it unlocked and close to your iPhone in order to unlock your iPhone.

How to Use Gestures

With Apple Watch, you can use gestures to perform different functions. Some of these gestures include:

Tap- Tap on the screen of your Apple Watch to select an item or a button. However, if your watch has the Always On feature, tapping the screen brings full brightness.

Tap and Hold- Tapping the screen and then holding your finger in place changes the watch face.

Drag- Perform this gesture if you want to adjust or even scroll a slider. To do this, just drag your finger across the screen.

Swipe- Use your finger to swipe up, down, right, or left for different screens to display.

How to Check Notifications

When a new notification comes to your Apple Watch, it will vibrate. This can be notifications of messages, activity reminders, noise alerts, or meeting invitations. Usually, your watch will display the notifications as they come in and you can view by raising your wrist.

Use the Digital Crown to scroll through the message. Once done reading, you can dismiss the message and tap Dismiss.

If you don't read the message right away, it will be saved for later in the Notifications Center. To view it:

Press and hold the screen and swipe down to open Notification Center. Alternatively, you can swipe down the Watch Face to open the Notification Center.

Turn the Digital Crown to go through the list of Notifications on your Apple Watch. Alternatively, you can swipe your screen up or down.

Tap the notification to view or respond to it.

How to Close Apps

To close apps on your Apple Watch, follow these steps:

Press the side button to see the Dock that contains a list of apps that you have used recently.

Once the list has appeared, swipe left on the ones you want to close and tap on the red X sign.

This will dismiss the app.

Working out with Apple Fitness+

Apple Fitness+ is a subscription service that allows you to choose from a collection of workouts led by experts. It can be used in conjunction with your Apple Watch and iPhone.

While you follow the workout on your iPhone, your screen will also display in-session metrics like calories burned and heart rate, which are captured by your Apple Watch.

Monitor Your Health

One of the Apple's in-built apps is Health and Activity, which comes with new features.

Here are some of the features of the iPhone 13

A Comprehensive Home Screen

Unlike the previous iOS 12 interface, the iOS interface has two tabs; Summary and Browse.

The Summary screen contains a lot of health and exercise data while the Browse screen contains health categories that you can track with the app including Activity, Body Measurements, Cycle Tracking, Hearing, Heart, Mindfulness, Nutrition, and more.

If you want to expand the list of activities displayed on the Browse screen, go to the Summary screen and click on the Edit link. Then, click on the All tab to see all the activities in one list.

More Highlights for Each Category

The Apple 13 comes with a Highlight section that provides more insight into your activities, including the statistics. In this section, you will see highlights for each category including your current metric and how it compares to the previous day or week.

Female Health Tracking

A new feature in iPhone 13 is the Female Health Tracking that allows women to track their menstrual cycles. The app, which is available by default, can predict when you will have your period and the associated symptoms.

Hearing Health

There is a new metric for hearing health in the iPhone 13 that lets you know whether you are listening to loud music. The app offers guidelines for the audio levels when using headphones.

Collecting Health Data

Your iPhone automatically captures and saves valuable health data which allows for easy monitoring. The type of data captured by the iPhone are the following:

Your iPhone has built-in sensors that captures the number of steps you take, your walking speed, etc.

The iPhone also captures valuable mobility metrics like walking asymmetry, double support time, etc.

Audio levels from any type of headphones connected to the iPhone are also automatically captured and stored.

Setting up a sleep schedule in the health app allows the iPhone to capture the periods when you're lying in bed and about to sleep.

You can also download any previous health records about your allergies, medical conditions, etc.

How to Share Health Information

Oftentimes, you may want to share your health information with close friends and family, caregiver, or health provider to get adequate support. The Apple Health app allows you to share your health information including your health trends.

To share your health information with others, open the Health app. Scroll down to Sharing at the bottom of the screen and click on it.

The app lets you share a summary of each topic in a secure way. Besides, you can stop sharing when you deem fit.

If you want to share the information with your health practitioner, sign in to the health system online portal and input your health system information. When the app connects to the health system's online portal, your health records will be downloaded so your doctor can see.

Workout **Class**

Apple Fitness+ is a fitness service you can use with your Apple Watch and your iPhone.

It offers 11 different workout types like HIIT, Yoga, Strength, and more. It also has guided meditation sessions.

Ask Siri

Siri is a smart assistant in Apple's devices that uses machine learning that answers questions and give suggestions to users. It is basically an app that controls the aspects of your life by saying a simple. "Hey Siri." Interestingly, this app can also recall recent conversations.

Siri does this while maintaining privacy and security. There are certain Siri aspects like showtimes, movie info, translations, sports statistics, dictionary, bookings, and restaurant figures.

Setting up Siri

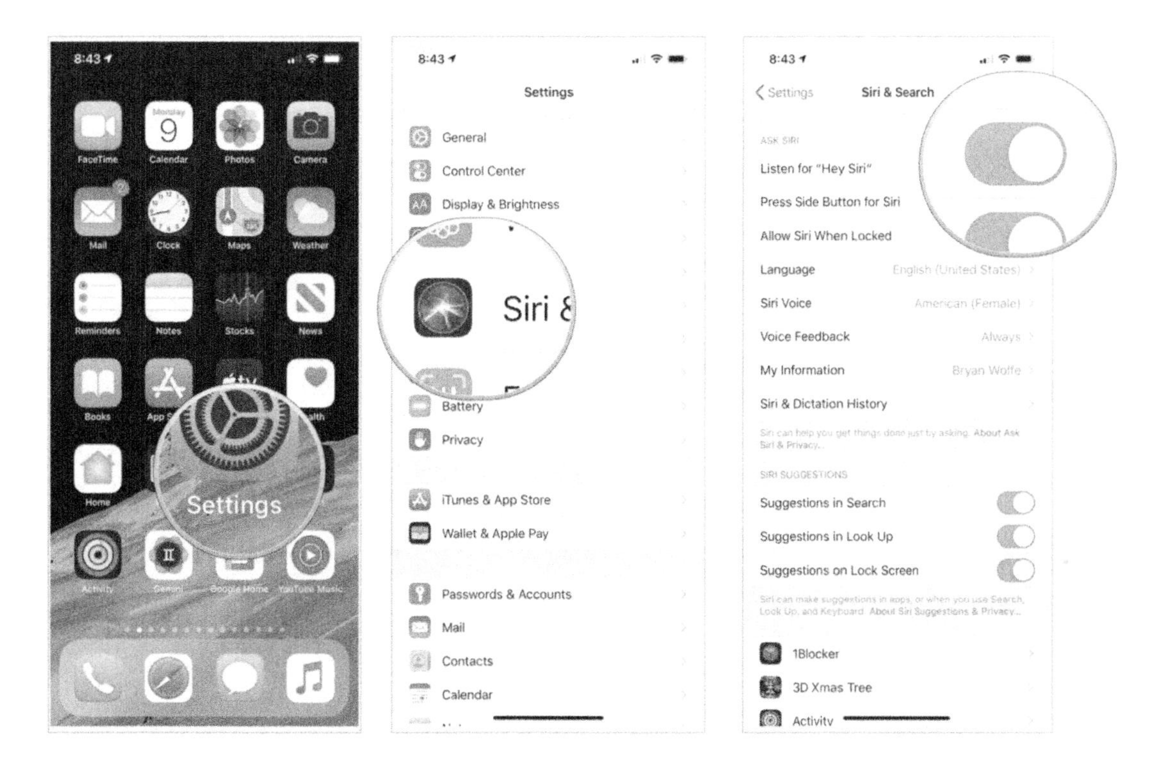

To set up Siri on your iPhone, go to Settings > Siri & Search, then do the following:

Turn on Listen for "Hey Siri" if you prefer to activate Siri with your voice.

If you prefer to have Siri activated with a button, turn on Press Side Button for Siri.

Activating Siri with your voice

Say "Hey Siri" out loud then ask Siri a question or order Siri to do a task for you.

• If you want Siri to answer another question or do another task, say “Hey Siri” again. You can also tap the icon.

Tip: You can prevent your iPhone from responding to “Hey Siri” by placing it face down.

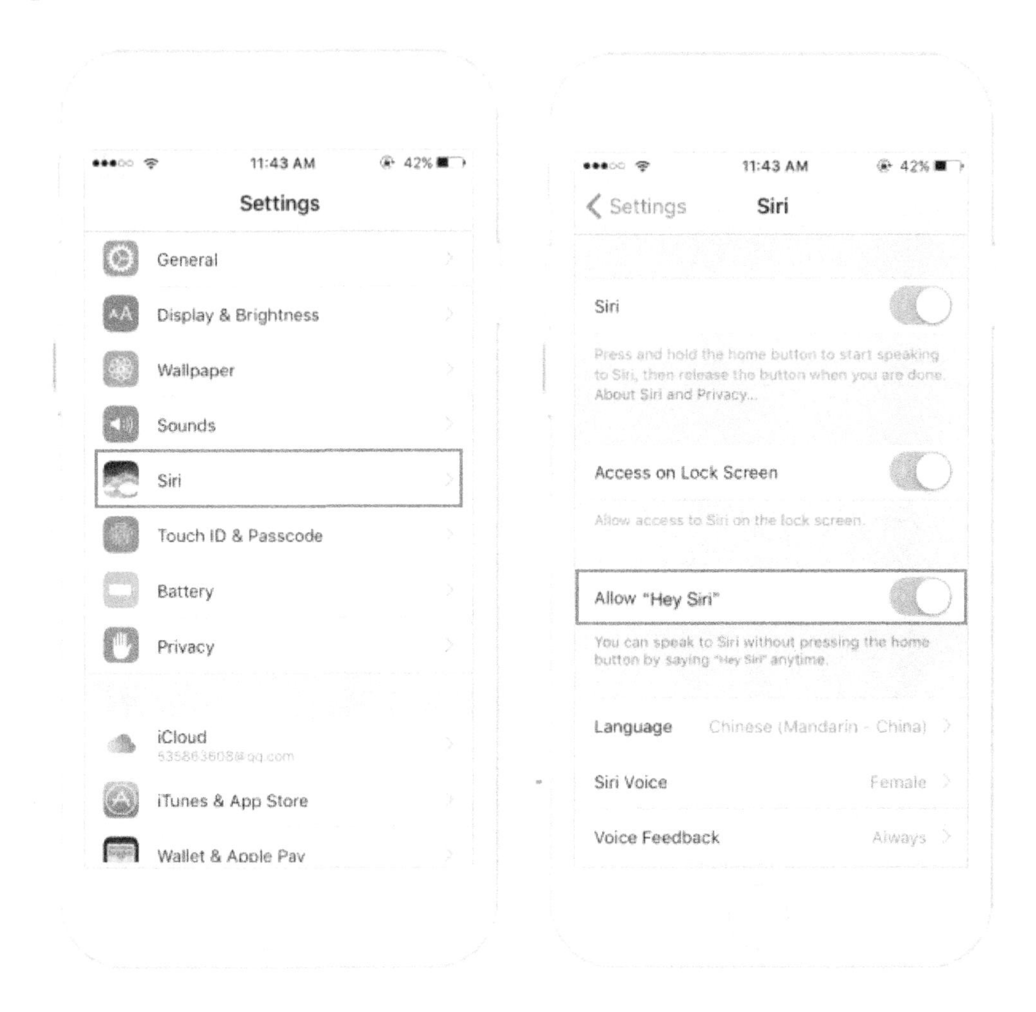

Activating Siri with a button

When Siri is activated with a button, Siri responds quietly when the iPhone is in Silent mode. Otherwise, Siri responds loudly.

Finding out what Siri can do

Using Siri to answer questions

You can use Siri to answer a lot of questions including, but not limited to:

- Checking facts
- Doing calculations
- Translating a phrase into another language

Using Siri to control apps

You can use Siri to do any of the following:

- Announcing calls and messages
- Playing music
- Using Maps and Maps widgets to get directions
- Controlling your home
- Adding shortcuts
- Sending messages and making calls

5. Using Siri in your car

You can keep your focus on the road by using CarPlay or Siri Eyes Free to allow Siri to make calls, send texts, play music or any do any other features on your iPhone.
CarPlay is only available in select cars, but it displays anything you want to do on your iPhone on the built-in screen. CarPlay also uses Siri so you can control CarPlay with your voice and activate Siri.
Siri Eyes Free, which is also available in select cars, allows you to control features of your iPhone with only your voice and not requiring you to look at or touch your iPhone.
Use Bluetooth to connect your iPhone to your car. Then, to activate Siri, press the voice command button found on your steering wheel and hold it until you hear the Siri tone. Once you hear the tone, you can then make a request to Siri.

The Settings of the Apps

The Settings app allows you to search for a specific iPhone setting you want to change, including any specific application settings.

To do this:

- Go to the Settings app, which can be found in the home screen or in the App Library.
- If you want to search for a particular setting you want to change, swipe down to show the search field then enter a term and tap the result.

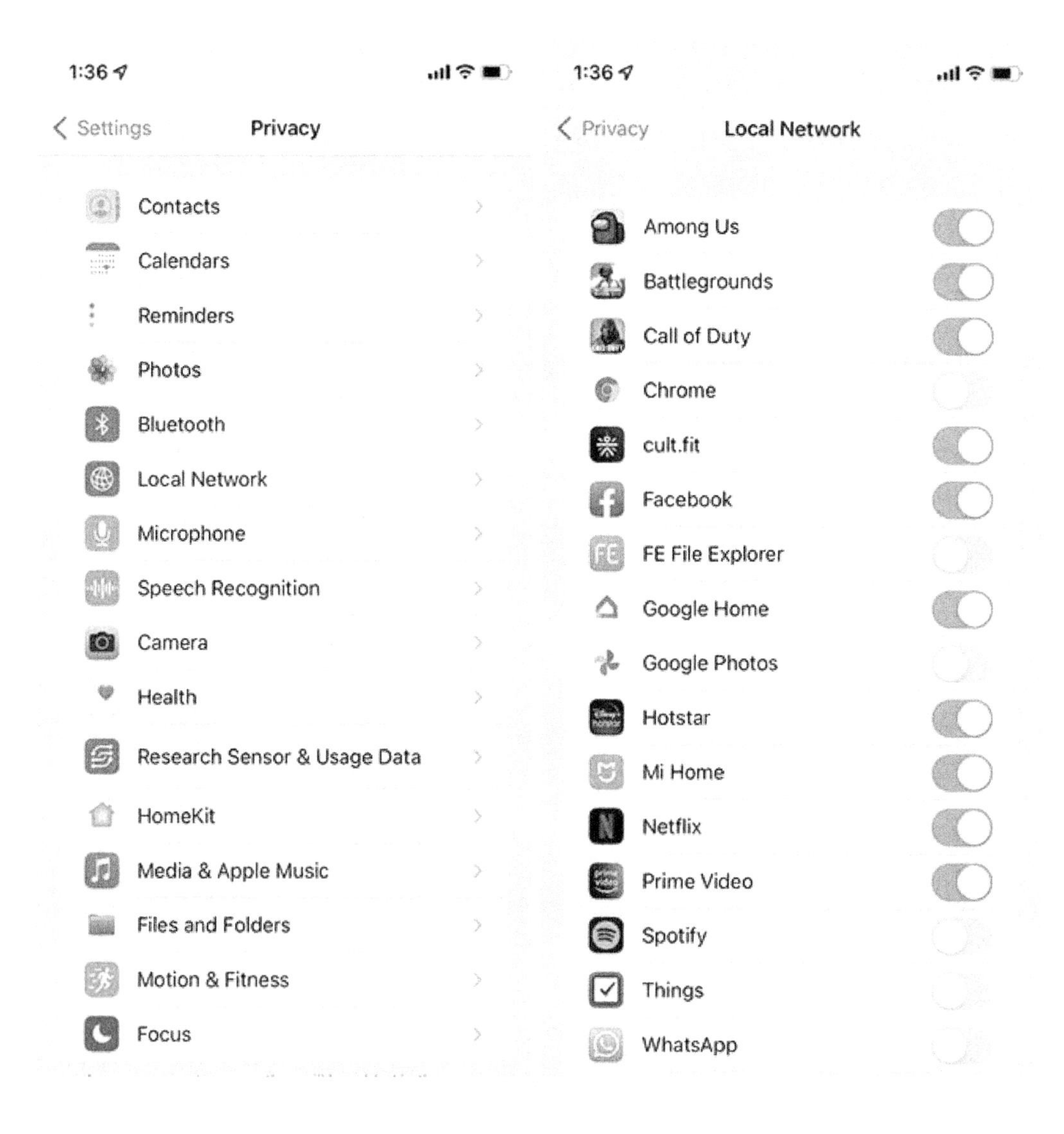

iCloud Setup and Setting

To set up your iCloud:

- Go to the Settings app.
- Select the Sign in to your iPhone option.
- Then enter your Apple ID and Apple password.

To synchronize your iCloud:

- Go to the Settings app and select your Apple ID. Then, select your iCloud ID.
- On each item, tap the switch button next to it to turn iCloud synchronization on or off.

Enabling automatic downloads

If you want to enable automatic downloads for your app, go to Settings and select App store. Tap the icon next to automatic downloads to toggle it on or off.
12.3 Update, Change, or Cancel your iCloud subscription

Upgrading your iCloud

If you want to upgrade your iCloud+ from your iPhone, do the following:

- Go to the Settings app. Then tap your name and go to iCloud > Manage storage.
- You can choose to Buy More Storage or Change Storage Plan.
- Then, choose a plan and follow the instructions that will be displayed on your screen.

Downgrading your iCloud

If your iCloud storage is more than you need, then you can choose to downgrade your subscription.

- Go to the Settings app then tap your name and go to iCloud storage.
- Tap Change Storage Plan.
- Select the Downgrade options and then enter your Apple ID password.
- After you've chosen a different plan, tap Done.

Cancelling your iCloud

If you want to cancel your iCloud subscriptions, do the same steps for downgrading your iCloud. When you get to the part where you have to choose a new plan, select the free 5GB one.

You'll still be able to access your current storage level until the billing cycle ends, then it will automatically switch to the free plan.

Finding Lost Devices on iPhone

There is nothing as heart-breaking as losing your iPhone. Fortunately, Apple provides the option to track and locate a stolen or lost phone. iPhone 13 is one of the iPhone models that comes with the tracking feature.

What's more, the new iOS 15 allows your phone to be located even when it's turned off or if the battery has run out. You can identify if the feature is active if the phrase "iPhone Findable After Power Off" shows up when you shut your phone down.

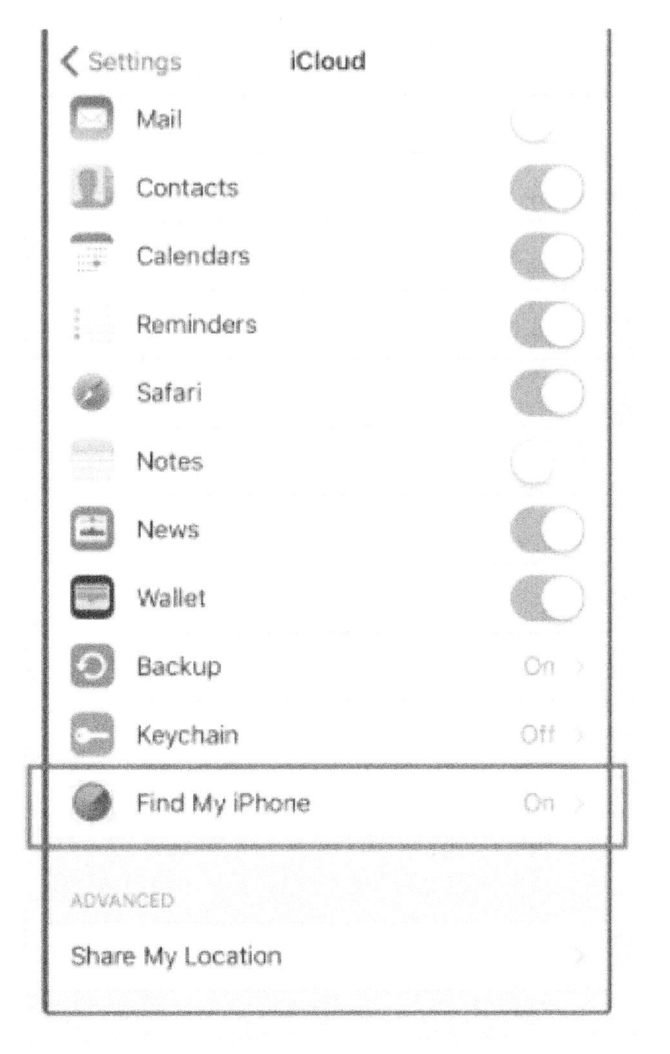

Activating the Find My app

If the feature to track your iPhone device is not active, here is how you can turn it on.

To set up this feature:

- Go to the Settings app, then tap your name and then tap "Find My."
- Select "Find my iPhone."
- Switch "Find my iPhone" on. You can also turn "Find my Network" on. This will allow you to find your phone even if it's not connected to an internet.
- Select "Send Last Location" to automatically send the location of your iPhone to your Apple account when your phone battery is about to run out.

How to See the Location of Your Device

You can use Find My App to see the current or last location of your device on the map. To do this:

- Scroll down to the bottom of the screen and tap on Devices.
- Locate the name of the device you want to find and click on it.
- The device will appear on the map if it is located.

- If "No Location found" appears on the device's name, then it means that the device cannot be found. In this case, go to Notifications and turn on Notify When Found to get notified once your device is located.

Enable Location Services

How to turn on Location Services

- First, go to the Settings app.
- Select Privacy and tap Location Services.
- Tap the switch right next to Location Services to turn it on or off.

Turning on Location Services for Applications

If you want to toggle Location Services on an app-by-app basis:

- Go to your Settings app.
- Tap Privacy and select Location Services.
- The screen will display the list of apps that want to use your location. Some apps want to use your location more than others. For example, other apps might have the Always or Never options, while the rest has the While Using the App or Never options.
- Scroll through the list of apps and enable Location Services for the apps you want to give your location to. If you want them to know exactly where you're located, you can turn on Precise Location.

Sharing Location Services with Family

To be able to do this, you need to set up Family Sharing first. Your family members will then need to share their location through Find My.

After that, you can do the following:

- Go to the Settings app and select Privacy.
- Select Location Services and tap Share My Location.
- Tap the switch next to Share My Location to turn it on.
- After this is enabled, you can use the Find My app to always know where you are based on the location of your device.
- If you want to stop sharing your location with a specific family member, tap that person's email address, then select Stop Sharing My Location.

CHAPTER 13: SOLVE COMMON PROBLEMS

All the phones in the iPhone 13 series encounter bugs and problems that can be frustrating when you don't know how to solve them. Fortunately, you may be able to fix some of these problems on your own, with a little help. This article will guide you on the possible iPhone 13 problems and how to fix them.

1. Heavy Battery Drain

A common issue that most iPhone 13 owners encounter is their battery draining fast. Although the iPhone 13 comes with a capacious battery, your battery life may run out quicker than you expected, hindering you from getting the most out of your phone.

How to Fix:

In this case, you may want to try any of the fixes below:

- Try restarting your phone first before doing anything else. Leave it turned off for at least a minute and then switch it back on. If the battery still drains quickly, then proceed to the steps below.
- You might want to try updating your iPhone. Upgrades usually involve fixing bugs so new firmware might have the potential to fix your battery issues.
- Turning off your 5G connectivity might also help. While it boasts a faster connection, 5G could drain your battery life much quicker than an LTE does.
- Consider switching off background refresh. To do this, go to **Settings** and tap on **General.** Click on **Background App Refresh** and toggle the button to turn it off.
- You can also check the performance of your apps and see how they're impacting your battery life. To do this, go to **Settings,** tap the **Battery** option and select the **Battery Usage** tool. Try deleting and then reinstalling any app that might be the cause of the abnormal drain.
- The other possible fix is to update your apps. Launch the **App Store** and navigate to the **Updates** section at the bottom of the screen. Locate **Update All** and tap it to update your apps.

- Contact Apple customer service if the above fixes do not resolve the issue. You can also take your device to an Apple store for the technicians to determine if your phone has a hardware issue that is causing your battery to drop.

2. iPhone 13 Charging Issues

Charging issues on iPhone 13 are mostly caused by the wireless charging feature. The battery percentage can get stuck during charging, making it impossible to charge your phone.

How to Fix:

- Try resetting your iPhone 13. To perform a reset, press the volume up button hold and do the same with the volume down button. Then, press the power button and hold it till your device powers off. Power on your phone and see if the issue is resolved.
- Update your iOS to make sure your iPhone 13 is using the latest stable version. Go to **Settings,** then tap **Software Update.** Download and install any updates that are available. If your iPhone 13 is up to date, then disregard this solution.
- Launch the **Settings** app and tap **Battery**. Locate and tap the **Battery Health** and then **Turn Off Optimized Battery Charging.** Contact Apple customer service to fix the issue.

If you're using a cable charger instead of the wireless option, then you can try any of the fixes below:

- The problem may stem from the charging cable. Check that it's plugged in properly. If the issue is still not resolved, you may want to replace your charging cable with a new one.
- Try cleaning the Lightning port of your iPhone with a clean toothbrush or anti-static brush. Dust and debris that have accumulated over time may be the cause of your charging problems. Then, try charging your phone again.
- Check the power source that your charging cable is plugged into. If you're using a power adaptor, you can try switching to a different one.
- For iPhones that won't charge past 80%, Apple recommends cooling the iPhone down. This means moving it to a much cooler location, as keeping it in a warm place while charging might heat up the iPhone too much and cause it to limit itself to only charging up to 80%.
- If the issues still persist, the problem might lie in your hardware. In this case, you can contact Apple customer service or take your iPhone to the nearest Apple store to have it repaired.

3. Overheating When Running Apps

The other common iPhone 13 problem is overheating during charging, when restoring an iCloud backup, or when running apps. When your phone is running hot, these operations will get paused and will only resume after it has cooled down. You will also notice a significant decrease in your iPhone's performance when it gets too hot.

Here are a few specific things that could cause your iPhone to overheat:

- Gaming, especially with video games that are resource-intensive and have high-end graphics
- Streaming for a significant amount of time can also be the cause of your iPhone overheating. When you stream longer, your iPhone has to constantly load content and keep the display working, which then causes it to generate more heat.
- Downloading huge apps while you are still using the device is also another reason behind your iPhone overheating.

How to Fix:

- If your iPhone 13 overheats during iCloud restoration, wait for a while as the operation pauses and then resumes back.
- Avoid prolonged exposure to the sun and, as much as possible, keep your iPhone in a cool and sheltered location.
- Try turning your brightness down. You can do this by going to the **Control Center** of your iPhone and toggle the brightness level. You can also go to **Settings** and then **Display & Brightness** to adjust your phone's brightness.
- Turn off your iPhone's Background App Refresh. Your iPhone constantly working in the background might be the cause of it overheating.
- Cellular data demands a lot of power from your iPhone so try to use WiFi instead, as much as possible, to keep your iPhone from overheating.
- You can also manage and prevent your iPhone from overheating by not spending too much time on games or videos that demand a lot of processing power. If possible, reduce the time spent multitasking applications that have too many demanding processes.
- Uninstall any applications that may not have been downloaded properly or any that keep crashing.
- Power off your phone and power it back on again.

- If your phone has a case, remove it to see if the overheating issue will be resolved.
- Try switching to Airplane mode.

4. 5G Not Working

5G is a new technology that is present in iPhone 13 mini, iPhone 13, iPhone 13 Pro, and iPhone 13 Pro Max. However, to enjoy this technology, you need to have a plan with your carrier.

Therefore, if you do not have a plan yet, you may encounter issues of 5G not working on your iPhone. Other users complain and report a drop in the 5G signal, making it nearly impossible to access fast 5G internet on their phones.

How to Fix:

Try any of the troubleshooting methods below to see if this issue to see if 5G signal will work on your device.

- Before you do anything else, try restarting your iPhone first. Most problems and bugs you encounter on your iPhone are usually resolved by doing this simple step. If not, proceed to the next solutions
- Make sure that 5G is enabled on your device. You can choose between two options when switching 5G on: **"5G On"** and **"5G Auto."** 5G On enables your phone to constantly use 5G even when the network provides 4G/LTE Speeds while 5G Auto only choose a 5G network when a better network connectivity is available, causing your iPhone to immediately switch to 4G/LTE when the 5G network becomes sluggish.
- Go to **Settings** > **General** > **Software Update** and check if downloading any recent updates will resolve the issue.
- Verify your plan. Check and confirm whether it includes 5G connectivity. Another cause could be that you've run out of data in your plan.
- Check with your network carrier whether they support 5G
- Also check for a carrier update by going to **Settings** > **General** > **Software Update**
- Check your SIM. It might not have been positioned properly, causing 5G to not work on your iPhone. Eject your SIM and then carefully reposition it.
- If you're using Dual SIM in your iPhone 13, both lines will automatically switch to 4G. You can either opt to use a single 5G SIM or make sure that both SIM cards have got support 5G networks.

- Ensure that Airplane mode is toggled off. Go to the **Control Center** and **tap Airplane Mode.** Wait for about four to five seconds before tapping it again to make sure it's switched off. You can also access this in **Settings > Airplane Mode.**
- When you're traveling, make sure that your carrier supports 5G roaming.
- If Low Power Mode is enabled, it could prevent your iPhone from using 5G. To disable this, go to **Settings > Battery** and switch it off. You can also tap the Low Power Mode icon in the **Control Center** to turn it off.
- Go to **Settings > General** then select **Transfer or Reset iPhone.** Tap the **Reset** option, and then select **Reset Network Settings** from the list that shows up. Wait a couple minutes before checking if the issue has been resolved.
- Contact your carrier or Apple customer service if none of the proposed solutions work.

5. Issues Unlocking iPhone 13 with Apple Watch

Apple provides the option for iOS users to unlock their phones using their Apple Watches. However, some users have reported receiving the error message that says "Unable to Communicate with Apple Watch" when they try using this feature.

How to Fix:

If you encounter this Apple bug issue, Apple recommends doing the following:

- Download and install iOS 15 Software Update on your iPhone 13 to resolve the problem. Go to **Settings**> **General**> **Software Update**.
- Download and install the Watch OS8 update.
- Restart your iPhone and Apple Watch.
- Unpair and try pairing your Apple Watch again.
- Check if your Apple Watch is actively connected to your iPhone 13. Go to the **Control Center** of your Apple Watch by swiping up the bottom of your screen. Check if the green iPhone icon is visible on the upper left of the screen. If not, enable your iPhone's Bluetooth and make sure it is connected your Apple Watch.
- Make sure that Unlock with Apple Watch is enabled on your iPhone. To toggle this, go to **Settings > Face ID & Passcode.**
- Go to the Apple Watch app on your iPhone. Select **Passcode** and then make sure that **Wrist Detection** has been switched on.

- Try resetting Face ID on your iPhone or creating an alternate Face ID.
- If nothing still works, contact Apple Support to resolve the issue.

6. Cellular Data Problems

In some instances, iPhone 13 models may display a random "No service" symbol, meaning that you can't connect to the internet. This also makes it possible to make and receive calls or even send texts. The No Service problem is likely to occur when your iPhone is switching between different carrier cells.

Issues with cellular data can also be caused by damaged hardware, basic software glitch, or a system-wide outage that causes the entire network to be unavailable.

How to Fix:

- Ensure that the area does not have a network outage.
- If there is a network in your area, try restarting your phone and check if the error has been resolved.
- Try and check if you've reached the limit on your Cellular Data plan. Go to **Settings** > **Cellular.** Scroll down further to see the status of your Mobile Data Usage.
- Turn on Airplane mode for several seconds and then disable it.
- Disable Data Roaming.
- Turn off your iPhone and then eject your SIM Card. Wait for about thirty seconds before re-inserting your SIM Card. You can also try another SIM Card with your iPhone to check whether the issue comes from the card or your iPhone.
- Go to **Settings** > **General** > **About** and see if there's any updates available. If there are, download and install them to update your Carrier settings.
- Shut off cellular data to see if the issue will be resolved. Open the **Settings app** and tap **Cellular**. Locate **Cellular Data** and toggle it off for about a minute before you toggle it on again.
- Reset your APN by going to **Settings** > **Cellular** > **Cellular Data Network** > **Reset Settings** and see if it resolves your Cellular Data problems.
- If you can't identify the problem and the issue still has not been resolved, contact your carrier or Apple customer service.

7. Unresponsive Touch Screen

According to some users, some of the iPhone 13 models are not responsive to tap. To make the touch screen work, one may be forced to restart their device. This issue appears to be a common issue on the upper corners of the screen.

How to Fix:

- When your iPhone 13 becomes unresponsive to taps, the quick way to fix it is to restart your iPhone.
- Make sure that Touch Accommodations is disabled. To do this, **activate Siri** then say, **"turn off Touch Accommodations."** You can also disable it on your iPhone 13 Pro Max by clicking the Side button three times.
- Remove your screen protector and then try if your iPhone responds to your taps.
- Since this is a software bug, updating your device to the latest iOS 15 version can help solve the issue.
- Another way to fix the software issue is by doing a factory reset. Make sure that you've backed up your iPhone before proceeding. If you can still navigate your screen in some way, go to **Settings** > **General** > **Transfer or Reset iPhone** > **Erase all Content and Settings.** If navigating your iPhone becomes impossible, you can plug it into your computer to perform the factory reset. Then, check if the problem has been resolved.
- If none of the listed solutions work, you might need a physical repair. Go to an authorized repair shop in your location and have your iPhone touchscreen fixed.

8. Face ID Problems

Face ID in Apple devices helps to provide additional security for your phone. However, you can encounter issues using this feature and you may be unable to unlock your phone with your face. Here are recommendations for fixing this issue:

How to Fix:

- Update your phone to the latest version of iOS.
- If the issue still persists even when your phone is running the latest version, you need to launch your Face ID settings. To do this, open the **Settings** app and scroll to **Face ID &**

Passcode. Enter your passcode and browse through the setup to ensure that all features useful for Face ID are turned on.

- After you have made Face ID changes, ensure that you look actively into the screen when using your face to gain access to your device. Bring the phone closer to your face so that it can register it. The maximum distance of your iPhone from your face should be around ten to twelve inches away.
- Also, ensure that your FaceTime camera is free from any debris. Also, the room should be well lit.
- Make sure that nothing is blocking your face. If you're wearing any sunglasses, remove them.
- If the Face ID issue is caused by your change of appearance, you need to set up an alternate appearance. For this one, go to **Settings** and tap **Face ID & Passcode**. Click on **Set Up an Alternate Appearance** to set up an alternate appearance to your Face ID.
- Try resetting your Face ID and then setting it up again.
- Make sure your phone battery is not below 10%. Having low battery could be the cause of your Face ID problems. If you have low power mode enabled, switch it off.
- Restart your iPhone and then check if your Face ID works.
- Consider resetting all the configuration settings in your iPhone and starting over from the beginning. Go to **Settings,** then tap **General** and select **Transfer or Reset iPhone.** Choose the **Reset** option and then **Reset All Settings.**
- If none of the solutions work, it might be a hardware issue. In this case, contact Apple customer service or go to the nearest Apple Store in your location to have your iPhone repaired.

9. Sound Problems

Do you encounter sound problems with your iPhone 13 speakers? Well, this is also a common issue where your phone's speakers do not produce clear audio. Fortunately, there are things you can do to get rid of this problem.

How to Fix:

- Try restarting your phone if the microphone starts to cut out or stops working.
- Verify that the SIM card tray is well-positioned as wrong positioning can cause sound distortion.

- Check the lighting port on the bottom of your iPhone and ensure that it is free from debris.
- If you are still encountering issues with your device's microphone, consider restoring from a backup.
- If you have problems hearing incoming calls and notifications, make sure your phone isn't on Silent Mode.
- Make sure that your iPhone isn't currently connected to any Bluetooth accessories like speakers or headphones. To disable your Bluetooth, go to your iPhone's **Control Center.**
- Try removing your iPhone case. It could be covering your iPhone's speaker holes.
- Disable Focus Mode as it could prevent you from hearing any sounds and alerts.
- Your iPhone could be stuck in Headphones Mode. Plug your headphones in and play any audio. Then remove the headphones and check if you can hear audio from the iPhone's speakers.
- Updating your iPhone might fix the bug. Go to **Settings** > **General** > **Software Update** to download and install any available updates.
- Go to **Settings** > **Sound Haptics** > **Ringers and Alerts** to test your iPhone speakers. You can increase the volume by moving the slider to the right. If you hear a sound, then your speakers still work. If not, then your iPhone might need a hardware repair.
- Contact Apple if the sound issue still persists and this may indicate hardware issues.

10. Apple Music Not Working

This was a problem when iPhone 13 was first launched and most users were not able to access Apple Music for those devices that were restored on iCloud backup. Because of this issue, it also becomes impossible to open the Apple Music Settings. What's more, Sync Library may not be available for your device. If you have experienced any of these issues, try these steps:

How to Fix:

- Make sure that you're connected to a stable internet connection.
- Launch the Settings app and open General.
- Click on Software Update.
- Here, you will see the Download option. Tap on it followed by Install or Install Now. Read and follow the on-screen instructions to get the minor update that Apple released to remove the bug.

- If it doesn't work, check the status of your Apple Music Subscription. Go to **Settings > iTunes & App Store > Apple ID.** Tap **View Apple ID** and then select **Subscriptions.** If your subscription has expired, it might be the cause of the issue.
- Try restarting your iPhone to remove any glitches in the software.
- Consider enabling iCloud Music Library by going to **Settings > Music > Enable iCloud Music Library.**
- It might be a server problem so check if the Apple Music Servers are still up or if they're experiencing some downtime.
- Try logging out and logging back in into your Apple ID account.
- The issue might be because of recently changed network settings. Go to **Settings > General > Reset Network Settings.**
- You can do a **Factory Reset** on your iPhone to fix the problem. Make sure your iPhone is backed up before proceeding.
- If nothing works, contact Apple customer service or visit the nearest Apple Store in your location.

11. Activation Problems

You can encounter some activation issues with your iPhone 13 during the setup. This can be caused by several things including when you don't have an active plan, or you do not run on the latest iOS version.

Other reasons that could cause activation problems with your iPhone 13 include:

- SIM carrier is temporarily down or inactive
- Minor bug in your iPhone's iOS
- SIM card is not inserted properly
- The iPhone 13 is connected to weak or poor internet
- Apple Activation services might be down due to maintenance

How to Fix:

- Check to ensure that the Apple systems in your phone are up and running. Go to Apple's System Status page to see if there is a green symbol next to iOS Device Activation. If not, then this could be the reason Apple's systems are not working as required, causing activation problems.

- If the services are working normally but you are still encountering activation issues, check to ensure that you are using the right SIM card and that it is correctly installed.
- However, if you have verified the SIM card but still get an alert that it is invalid, do the following.
- Restart your iPhone.
- Update your device with the latest iOS version.
- Check your plan and ensure that it is active.
- Confirm the carrier's settings update by: Launch **Settings** and browse to **General.** Tap **About** to get any available updates.
- If there is, click **Update** or **Ok**.
- You can also check if your SIM card is inserted correctly as it might be causing the issue.
- Make sure that your iPhone is connected to a stable and good internet before activating it.
- You can use iTunes to activate your iPhone 13 by turning off your iPhone and plugging it in to your computer. After launching iTunes, turn on your iPhone. An **"Activate your iPhone"** window will pop up on screen once iTunes detects your iPhone. Enter your Apple ID and passcode then select **Continue** to activate your iPhone.
- If you are still unable to activate your iPhone 13, contact your carrier or get in touch with Apple Support customer service.

12. Automatic Shift Between Wide Cameras and Ultrawide

This is a common issue in iPhone Pro, which features a new macro mode that is designed to work with the 12MP ultrawide camera. However, some people report an issue with this since the camera automatically switches the lens, from the primary ones to the ultrawide ones.

While Apple says that this is a new feature for this model, it is an issue for some since the frame captured is different from what is in the viewfinder.

How to Fix:

- Make sure your iPhone 13 is running iOS 15.1 or higher. If not, you can go to **Settings** > **General** > **Software Update** to download and install the latest updates.

- Go to **Settings,** then select **Camera.**
- Toggle the **Auto Macro** switch off to disable automatic macro mode. If you want to take pictures in macro mode, you can still take them manually.

13. Wi-Fi Issues

Just like cellular data issues, Wi-Fi issues are also common in iPhone 13 series. This occurs when your Wi-Fi is slow or you have dropped connections.

How to Fix:

- Start by ensuring that your Wi-Fi network is functioning and there is no network outage. You can restart the router to solve any network issues.
- If your router is functioning and there is no network outage in your area, the next step is to tweak your phone's settings. You want to forget the Wi-Fi network and connect to it again. To do this:
- Go to Settings and click Wi-Fi.
- Locate your connection and tap (i) in the circle.
- At the top of the screen, you will see the option to forget the network. Make sure that you have the network's password before performing this step because you will need it to log back.
- You can also try going to **Settings** and then **Wi-Fi** and toggling the switch off and on.
- Consider restarting or rebooting your iPhone if the issue still persists.
- If you have VPN enabled, go to **Settings** and turn it off. Then, try to reconnect to your Wi-Fi network.
- Downloading the latest iOS updates to your iPhone might also help as pending updates can cause problems in your Wi-Fi connectivity.
- Make sure that your date and time is set correctly as this could cause issues with your iPhone, including Wi-Fi connectivity issues.

If this does not work, the next possible fix is to reset your phone's network settings. Here is how:

- Open the **Settings** app then **General**.
- Click on **Reset** and tap **Reset Network Settings**.

If nothing works, contact your internet provider or get in touch with Apple Support as your iPhone may have a hardware issue.

14. CarPlay Issues

Some users have reported that playing music via CarPlay makes the system crash while others noted that it doesn't play music on iPhone 13 Pro.

There could be different factors that cause these problems with CarPlay. These include:

- Your iPhone restricting the Apple CarPlay application
- Faulty wireless or USB connection in your car
- If Siri is not enabled or has issues, this could result in your Apple CarPlay malfunctioning
- Glitches in iOS upgrades might interfere with the CarPlay system

How to Fix:

To fix this issue, here is what other users recommend:

- Reset all network settings.
- The other possible fix is to disable the equalizer.
- Turn off the "**Late Night**" toggle in the Apple Music Settings.
- Try disabling the Dolby Atmos feature by opening your iPhone and navigating to **Settings** > **Music** > **Dolby Atmos.** You can either turn it off or set it to automatic. Then, try restarting your iPhone and reconnecting to CarPlay.
- Make sure you've allowed screen time for your CarPlay. Go to **Settings** then select **Screen Time** if it hasn't been enabled. Then, go to **Content & Privacy Restrictions.** It will prompt you to enter your password. Go to **Allowed Apps** and **Enable CarPlay.** Restart your iPhone and apply the changes immediately.
- Go to **Settings** > **Siri and Search** and make sure the **Listen for "Hey Siri"** option is enabled. Then toggle the **Press Side Button for Siri** and **Allow Siri When Locked** switches on.
- Try turning off sound recognition by going to **Settings** > **Accessibility.** Tap on the **Sound Recognition** option and disable it.
- Consider refreshing Airplane mode to correct connectivity issues in your iPhone. Go to **Settings** to **Enable Airplane Mode.** Then, restart your iPhone. When your iPhone turns back on, switch off Airplane Mode.

- Make sure your iPhone is properly connected. You can try changing USB cables or ports, just in case your current one might be faulty.
- You can also disable Bluetooth on your iPhone and then restart the device. Reconnect to CarPlay again when it's back on and see if it works.
- Alternatively, resetting your CarPlay connection or getting rid of older Bluetooth connections to prevent them from interfering with your CarPlay connection might help.

If all else fails, try resetting your iPhone settings:

- Go to **Settings > General.**
- Select **Transfer or Reset iPhone.**
- Tap **Reset > Reset All Settings.**
- Enter your device passcode then confirm the process.
- After it completes, make sure to restart your iPhone.
- Then, connect to CarPlay again and see if the problem has been fixed.

If the problem still persists, reach out to Apple Support.

15. WhatsApp Mailbox Keyboard Flickering/Flash

After upgrading to iOS 15.0.2, many users find that mailbox and WhatsApp keyboards keep flickering or flashing. The official support on a discussion. Apple points out its iPhone hardware to blame. However, it doesn't make sense. A high chance still goes to iOS 15.0.2 bug. So a tenable solution is waiting for the official fix, basically on the next iOS update. Or downgrade iOS 15 to iOS 14 on your iPhone 12 Pro/Max or older iPhone.

16. The Speaker Volume Didn't Work On Youtube, Phone Calls

The latest gripes about iOS 15.0.2 upgrades also cover no volume or smaller volume on YouTube videos, even phone calls. Restart your iPhone twice. It's proven effective, though the volume is still smaller than before.

17. Unable To Update To iOS 15.x

Two workarounds are provided for discussion. Apple by a zealous iPhone user. First of all, the range of the mobile network is minimal. Do not wait until 'No service' appears. Secondly, check if your iPhone is connected to the Internet through the Personal Hotspot of another iOS device. If so, switch to your cellular data or Wi-Fi network.

18. iOS 15 Stuck On Verifying Update

As one of the most common errors, many users have iOS 15 frozen on Verifying update screen on iPhone devices during the installation and have their devices unusable. This is annoying, but the good news is that there are many hidden iOS 15 tips and tricks to iOS 15 problems of this kind. iOS 15 update freezing problem troubleshooting:

- Give it more time;
- Delete the update, restart iPhone and try again;
- Factory reset and redownload;
- Do a hard reset or hard reboot to your iPhone.

19. Insufficient Space For iOS 15 Download

You only need 4GB of free space on your iPhone. This is not a big deal for 64 or 128GB iPhone users. For 16 & 32GB users, you might face the common iOS 15 update problem: insufficient space for the download. To solve this issue, you can try:

- Click "Allow App Deletion" if you don't use those apps often;
- Delete unwanted videos, music, photos, etc. instead of apps;
- To backup iPhone and delete files if you want an iPhone backup.
- Check how much space is left > Rent or purchase an HD movie in iTunes App (exceed your free space) > Click Rent or Purchase, and iPhone will help auto clean up space > Cancel rent/purchase and check your iPhone space now (much more space available).

20. Data Loss After iOS 15 Update

Another problem that most users, if not all, often ignore is the iPhone data loss problem during the process. They can't instantly discover this issue until they need the file someday. Certainly, you do not need to be panic if you have iPhone backups on your computer or iCloud. What if you don't constantly backup your iPhone? Try the below tips to fix data lost after the iOS update. Tackle iOS 15 Update Problem with Data Loss:

- Check if you have an iPhone backup with iTunes. If yes, restore iPhone backup with iTunes.
- If you don't constantly update the latest iPhone files to iCloud, just restore the iCloud backup to iPhone to get as much data you can. Minimize the data loss.
- Check if you can recover some data via your social media accounts, e.g., Instagram, FB, Twitter.
- Log in with your other Apple ID accounts to restore files if you rent or purchase with different Apple IDs.

21. iOS 15 Stuck On Recovery Mode

Sometimes, the iPad, iPhone 11, and older iPhones are stuck into a recovery mode that seems endless. If it freezes your iPhone, you're not alone. Take it easy; here are some useful fixes and hacks to problems and get your device out of recovery mode.

Solution:

- Try to do it via iTunes if you're updating OTA;
- Disconnect iPhone to computer and turn it off. Then reconnect to the computer and hold the home button until iTunes detect your device;
- Make sure you have free up enough space on iPhone;
- Uninstall and reinstall iTunes;
- Download redsnow even if your iPhone is not jailbroken.

22. iOS 15 Stuck On Apple Logo

Many posts on the Apple forum about iOS 15 update errors with a stuck Apple logo screen. If you have tried hard resetting your iPhone, plugging into iTunes, but the Apple logo continues on the iPhone screen, let's solve this common issue below.

Problems And Fixes:

- Reset your stuck iPhone to Factory settings;
- Try the TinyUmbrellla fix recovery app;
- Try to use the SSH method (remove DYLD_INSERT_LIBRARIES file and restart iPhone);
- Try a DFU (default firmware update) restore, which can be complicated;

- Try to use RedSnow, which requires you to jailbreak your iPhone.

23. iDevices Get Bricked After iOS 15 Update

When iOS 14 was released, massive users have suffered the iOS update bricking iPhones problem. Unfortunately, when updated to iOS 15, it still happens to a small number of users. Fortunately, the effective fixes and tricks are ready for a try.

Fix Problems:

- Upgrade to iOS 15 or restore your iPhone with iTunes;
- Sign in to iCloud and remove Activation Lock;
- Perform a factory reset to your iPhone or iPad.

24. iOS 15 Battery Drains Fast

Many users run into fast battery drain on iOS updated iPhone iPad. Even on the latest iOS 15, users still have iPhone battery troubles. iOS 15 is a battery killer. An iPhone X of 100% battery drops to 46% without upgrading to iOS 15. And another user reported that his iOS 15 iPhone 11 battery jumped from 97% to 80% within half an hour at a lock screen state.

Solve Problems With Battery Drain:

- Always reboot your iPhone;
- Turn off Background App Refresh by going to Settings > General > Background App Refresh;
- Disable the iCloud Keychain at Settings > iCloud > Keychain;
- Set the iPhone back to factory setting;
- Go into Settings > Battery and turn off apps that kill the battery life;
- Reset your Network Settings;

25. iPhone Is Overheating After iOS 15 Update

iOS 15 update also causes overheating issues on iPhone 12/11 or older models. Some users experience extreme overheat problems, while others have their iPhone heat up when it starts to play video, download files, or play music. It's often a bad battery that should be responsible for overheating issues, but besides that, here are some other tips.

Problems And Fixes:

- Resort to iTunes recovery if iPhone gets hot;
- Check if there is any wonky widget and disable that widget;
- Sign out your iCloud account if necessary;
- Disable the automatic update of the apps, and remove unnecessary widgets;
- Reset your iPhone.

26. iPhone Running Slow After Updating to iOS 15

Almost every iPhone iPad user complains about iOS 15 glitches with slow performance. And many users talk about this problem on YouTube and point out that Apple might use the iOS 15 updates to slow down older iPhones on purpose. Purposely making them slower is to force iPhone owners to update to the new iPhones or lose app compatibility. Other iPhone users think that Apple downgrades users' current iPhone iPad to make the newly released iPhone 11 look faster. No matter what they analyze is right or wrong, the truth is that your iPhone is running slow on iOS 15.

Fix Problems:

- Upgrade to the latest iOS 15 as the update is designed to address the throttling issues.
- Downgrade. iPhone users all think iOS 15 betas now aren't so stable.
- Upgrade iPhone to newer iPhone with better hardware configuration.
- Patiently wait for the solutions from Apple or just wait for the release of the iOS 15 higher version.

27. iOS 15 GPS Problem

After iOS 15 upgrading, many iOS users have complained about GPS not working properly when using apps that require your location, such as Google Maps or Apple Maps. Within Google Maps, for example, the bug means you won't be able to get proper turn-by-turn directions, which can be a real drag on your trip. Worse still, other issues are either no GPS signal or slowly updating location signals.

- Reset Network Settings (Settings > General > Reset > Reset Network Settings)
- Reset Location & Privacy (Settings > General > Reset > Reset Location & Privacy)

- Ensure Location Services are ON (Settings > Privacy > Location Services > ON)
- Restore from iTunes backup: restore your iPhone in iTunes with your iOS 14 back-ups.

28. iOS 15 update Errors With Wi-Fi

"My Wi-Fi keeps dropping out; any suggestions?" This problem creeps up every year, and unfortunately, it's no different with iOS 15 betas now. Yes, even the latest iOS has trouble connecting secure Wi-Fi networks. Since iOS 15 was released, users have complained about Wi-Fi not working problems on the iOS 15 iPhone or Wi-Fi keeps disconnecting randomly. So how to fix the iOS 15 upgrade issues? (You can also try the fixes at the beginning of this part)

Ios 15 Update Problems Troubleshooting:

- Go to Settings > General > Reset > Reset Network Settings;
- Switch to Google's DNS: Settings > Wi-Fi> Click on the networknetwork>Delete all numbers under DNS and enter 8.8.8.8 or 8.8.4.4.
- Delete any VPN app on your iPhone;
- Switch on airplane mode, turn Wi-Fi on, restart iPhone, and turn off AirPlane mode.

29. iOS 15 Problems With Cellular Data

Users are encountering abnormal cellular data problems on their devices, including Cellular data not working, the cellular data network is missing, and excessive data usage. Similar to Wi-Fi & Bluetooth not working on iOS 15 iPhone issues, it's hard to nail down the causes for cellular data errors on iOS 15, but there are some hidden tricks and hacks.

Fixes:

- Turn off Wi-Fi assist;
- Turn off LTE Voice, go to Settings > Cellular > Cellular Data Options > Enable LTE (turn it off), then under Roaming, turn off Voice Roaming, Data Roaming, and International CDMA;
- Turn "Use Cellular Data" for iCloud off;
- Turn off cell data for app updates;
- Take out the SIM card and put it back in.s;

30. iOS 15 App Crashing Constantly

After upgrading to iOS 15 (beta), the Apple forum has some people complaining about freezing or crashing apps on iPhones. Those frozen/crashed apps include FaceBook, Phone, Map, iMessage, Notes, and some more. The apps will be unresponsive for a few minutes. Apps like Notifications, Skype for IOS crash and stop working. The tips below may fix the problems with iOS 15 update crashing/freezing apps.

Fixes:

- Transfer contacts stored on Exchange server to iCloud;
- Disable the Exchange contacts;
- Perform a Reset Network Settings;
- Delete the App and Reinstall.

31. iTunes Errors With iOS 15 Update

As iTunes plays an important role, users will inevitably encounter issues with iTunes. The problem is myriad. There are iTunes Error 9006/1671, iTunes Error 14/39/54/3914... All those iTunes not working errors prevent users from upgrading iPhone iPad smoothly.

But Here Are Tips, Tricks, And Fixes:

- Check the security system of your computer and temporarily disable it if necessary;
- Install the latest version of Xcode 8 on Mac;
- Clear iPhone/iPad cookies, caches, and histories;
- Update iTunes to the latest version.

32. Camera Not Working Bug

IOS 15 comes with the newest and advanced camera effects, allowing you to add personality on FaceTime and iMessage. Yet it's not always good. There is a camera bug after upgrading to iOS 15: The camera is not working and always shows a black screen after people turn it on.

Troubleshooting:

- Force close the Camera App

- Restart your iPhone iPad of iOS 15
- Reset all settings. Go to Settings > General > Reset > Reset All Settings.

33. iOS 15 iMessage Issue

Some users find their texts are not getting sent and delivered on the iOS 15 iPhone. And some others tell us that the update changes the way Messages sends texts, sending out messages from email addresses rather than phone numbers.

Quick Tips:

- Turn off your iPhone or iPad. Remove your SIM card, readjust its seating, and reinsert it.
- Restart the Messages app by double-tapping the Home button or swiping up the Home Gesture Bar and then swiping up to the app to force quit. Once closed, relaunch the Messages app.
- Toggle iMessage off, wait 30 seconds, and toggle back on. Go to Settings > Messages > iMessage
- Go to Settings > Messages > Send & Receive and make sure your phone number AND Apple ID's email address (or iCloud email) is present and is checked under "you can receive messages to and reply from."

34. iOS 15 App Not Update

It's annoying when you need the iPhone iOS 15 apps to scramble for tickets to Taylor Swift concert or Comic-Con but find the app needs to update. What makes you even more off-putting is that your iPhone won't update apps after updating to iOS 15. Sad, angry, or furious? Don't be. Follow the hidden tips and tricks for iPhone to solve this iOS 15 update problem.

- Check your Wi-Fi network connection.
- Free up iPhone space with iOS 15 new features for App updates.
- Turn on Updates and Allow on iPhone settings.
- Remove outdated apps from your iPhone and try again.

35. Touchscreen Not Working On iOS 15

Users report that the iPhone's touchscreen is responding too slow, keeps flickering, freezes, or becomes unresponsive. There are some secret iOS 15 tricks when there is a touchscreen problem on iOS 15. See more about iPhone Touch ID not working problem

Fix problems:

- If you're using the 3D touch screen, first make sure you have turned on the 3D Touch option: Settings > General > Accessibility > 3D Touch;
- Restart or Force Restart your device to fix any possible internal errors;
- Delete and reinstall any problematic apps that may lead to unresponsive touch screen problems after the iOS 15 update;
- Free up iOS storage as a too-full iPhone will tend to run slower.

36. No 32-Bit Apps On iOS 15

It's reported that over 0.18 million iPhone apps won't be compatible with iOS 14. The same applies to iOS 15. Thus, after you update iOS 15 on iPhone, you need to carve out some time to replace 32-bit apps with 64-bit free iPhone apps. That might be time-consuming. But you have no choice. After all, you can't always enjoy iOS 15 new features but not bear its downsides.

Download And Install 64-Bit Apps On Ios 15 Iphone Ipad:

- Delete 32-bit apps and then open App Store.
- Type the app name or keywords to search the app you need.
- Click "Download" and enter your Apple ID.
- Organize the apps on your iPhone desktop using different folders.

37. Do Not Disturb While Driving Not Working On iPhone

The Centers for Disease Control and Prevention report shows that more people die in car crashes each year in the United States than in other high-income countries. There are many reasons to cause tragedies. Among them, distraction in the car is the prominent one. To improve this situation, Apple makes efforts to its new iOS 15 with a new feature, Do Not Disturb While Driving. The features of

DND while driving are good. Yet, some users sometimes find this feature not working on their iOS 15 iPhone. To solve this iOS 15 upgrade problem, you can try:

- Go to Settings to disable this feature. Reboot your iPhone and go to settings to turn it on again.
- Try to 3D Touch this feature on your Control Center to launch this feature.
- Give your iOS 15 iPhone a hard reset.
- Update iOS 15 betas to iOS 15 (latest version) to fix this unstable bug.

38. Notifications Freeze On iOS 15

An iPhone user complains on Reddit, and the notifications mess up the iPhone wallpaper. As you can see from the right picture, notifications are stuck, and eventually, they become part of the iPhone wallpaper. That's unaesthetic. Currently, there are no official solutions given by Apple, so just try the below iOS 15 secret tips and tricks:

Solutions To Ios 15 Update Problem In Notifications:

- Shut down your iPhone and restart it later.
- Head to Settings to change iPhone wallpaper to have a try.
- Turn off iPhone; pull the SIM card out of iPhone; restart your iPhone several hours later.
- Check if your iPhone is overheating. If yes, cool down iPhone and later use to have a try.
- Downgrade iOS 15 to iOS 14.

CHAPTER 14: FAQS

1. Does iPhone 13 support dual SIM?

Yes, all the iPhone 13 models support dual SIM with two active eSIMS or with a nano-SIM and an eSIM. Unfortunately, any existing micro-SIM cards are not compatible with iPhone 13 and iPhone 13 mini. You can use the two lines to send and receive messages on SMS, MMS, and iMessage. You can also use them to make and receive FaceTime calls and audio. However, it is worth noting that your iPhone 13 can only utilize one cellular data at any given time.

Does iPhone 13 have better battery life than iPhone 12?

The battery is one of the new features in the iPhone 13 models and is considerably better than that of the iPhone 12 models. Some notable improvements include a greater capacity of up to 10 to 15%, more power-efficient displays, physically larger in size, and considerably longer lasting battery life. The iPhone 13 mini and iPhone 13 Pro can last up to 1.5 hours longer than the iPhone 12. Meanwhile, the iPhone 13 and iPhone 13 Pro Max can last up to 2.5 hours longer than their iPhone 12 counterparts.

Is iPhone 13 Waterproof?

Yes, iPhone 13 features IP68 dust and a water-resistant rating. Both the dustproofing and the waterproofing on iPhone 13 can go up to 6 meters deep and it can be immersed for a maximum time of 30 minutes. All iPhone 13 models are then safe from any rain, accidental spills and any more water-related incidents.

What are the major differences between the iPhone 13 models?

Although these models share some similarities, they have their differences in terms of capacity, size, battery life, and more. Both the iPhone 13 and iPhone 13 mini share similar features except for their screen sizes. iPhone 13 has a larger screen size of 6.1" compared to the 5.4" screen size of the iPhone 13 Mini.

The other major difference is their resolution, with the iPhone 13 featuring a 2532-by-1170 pixel resolution while that of the iPhone 13 mini is 2340-by-1080 pixel. The display of both the iPhone Pro and the iPhone Pro Max is also one of the major differences between these high-end models.

iPhone 13 Pro Max has a display size of 6.7" and a resolution of 2778-by-1284 pixels. On the other hand, iPhone 13 Pro screen size measures 6.1" and the resolution is 2532-by-1170 pixels.

Another major difference between iPhone 13 and iPhone 13 Pro is the setup of their rear cameras. iPhone 13 has a dual 12MP camera system that comes Wide and Ultra Wide lenses. Meanwhile, the iPhone 13 Pro boasts a triple camera system that adds a Telephoto lens.

The storage options between the four models also vary. The standard iPhone 13 and the iPhone 13 Mini come in 128GB, 256GB, and 512GB storage options, while the iPhone Pro and the iPhone Pro Max also have a 1TB storage option, which marks the first time of having a 1TB storage option for an iPhone.

Additionally, only the iPhone 13 Pro and iPhone 13 Pro Max feature 120Hz refresh rate display while the standard iPhone 13 and iPhone 13 Mini models do not.

What should I do when Siri is not working on my iPhone 13?

Siri is a useful virtual assistant that can help you accomplish many tasks on your device without being hands-on. If your Siri app is not working, there are several things that you can do to get it fixed.

- Tap **Settings** on the Home screen.
- Navigate to Siri & Search.
- From the options provided, ensure that the three top options are enabled.
- If not, toggle their switches to turn them on.
- Restart your phone and check if the issue still persists.

Besides enabling Siri Settings, other quick fixes for Siri issues in iPhone 13 include:

- Update your phone
- Reset your phone's settings
- Ensure your iPhone device is facing upward.
- Confirm Siri language
- Ensure your internet connection is stable

Does iPhone 13 include an adapter?

No, you will have to purchase an adapter separately since it is not included with your device. However, what is included in the box is a USB-C cable.

How do I charge my iPhone 13?

You can charge your iPhone 13 with a MagSafe charger. This charger allows for wireless charging of up to 15W. It has magnets that can perfectly align with your phone, making charging fast and easy.

Is the MagSafe Experience on iPhone 13 different from that on iPhone `12?

Besides, the MagSafe capability for iPhone 13 is similar to that of the iPhone 12 series, including the maximum charging speed. However, iPhone 13 models can charge at 15 watts while iPhone 12 models charge at 12 watts over MagSafe.

What are the prices for the iPhone 13 Series?

Apple has priced each model differently depending on whether it is a level-entry or high-end model. Also, the price for iPhone 13 models varies with the storage. Here is a breakdown:

- **iPhone 13 Mini**- $729 for models with 128GB, $829 for models with 256GB, and $ 1,028 for models with 512GB.
- **iPhone 13**- $829 for models with 128GB, $929 for models with 256GB, and $1,129 for models with 512 GB.
- **IPhone 13 Pro**- $999 for models with 128GB, $1,099 for models with 256GB, $1,299 for models with 512GB, and $1,499 for a model with 1TB.
- **iPhone 13 Pro Max**- $1,099 for models with 128GB, $1,199 for models with 256GB, $1,399 for models with 512GB, and $1,599 for models with 1TB.

It is worth knowing that Apple charges $30 less if you activate your non-Pro models through a carrier.

Can I use my iPhone 12's case on my iPhone 13?

No, it is impossible to interchange and use your iPhone 12's case on your iPhone 13. This is because the iPhone 13 models are a bit wider compared to their predecessors. So you will need a new iPhone 13 case that will perfectly fit your device.

Besides, each iPhone 13 model needs its own case and they cannot be used interchangeably since they have different camera modules. This is true for the Pro versions, which are significantly bigger.

What camera features are included in iPhone 13?

iPhone 13 is equipped with a 12MP camera that comes with a wide camera lens that measures 26mm. Some of the camera features on this iPhone device include:

- Smart HDR 4
- Ultra-wide camera
- Auto-image stabilization
- Digital zoom (up to 5x)
- Sensor shift optical image stabilization
- Deep fusion
- Burst mode
- Photographic styles

What can I use Cinematic Mode for?

Cinematic mode is a new feature that has been included in all the 4 models of the iPhone 13. Simply put, it is a video mode that functions more like a portrait mode, allowing you to focus on the subject when shooting a video.

This feature can automatically change the subject in focus after shooting the video. Besides, you have the option to do this manually by tapping on the subject that is in focus. Generally, this is a good feature if you want to add a more dramatic cinematic feel to your video shots.

Do iPhone 13 models come with a lightning port?

Yes, iPhone 13 Mini, iPhone 13, iPhone 13 Pro, and iPhone 13 Pro Max all feature a lightning port. You can use this to charge your phone faster and even transfer data. Nevertheless, the iPhone is highly compatible with MagSafe charging that lets you charge your iPhone 13 wirelessly.

What is the display refresh rate of the iPhone 13?

Among the exciting features and upgrades on iPhone 13 is the display refresh rate. Both the iPhone 13 Pro and the iPhone 13 Pro Max support a 120Hz display refresh rate, which enhances the scrolling experience and makes games more responsive.

However, this new feature is not available on the iPhone 13 Mini and iPhone 13. Instead, these two models have the conventional refresh rate of 60Hz.

Printed in Great Britain
by Amazon